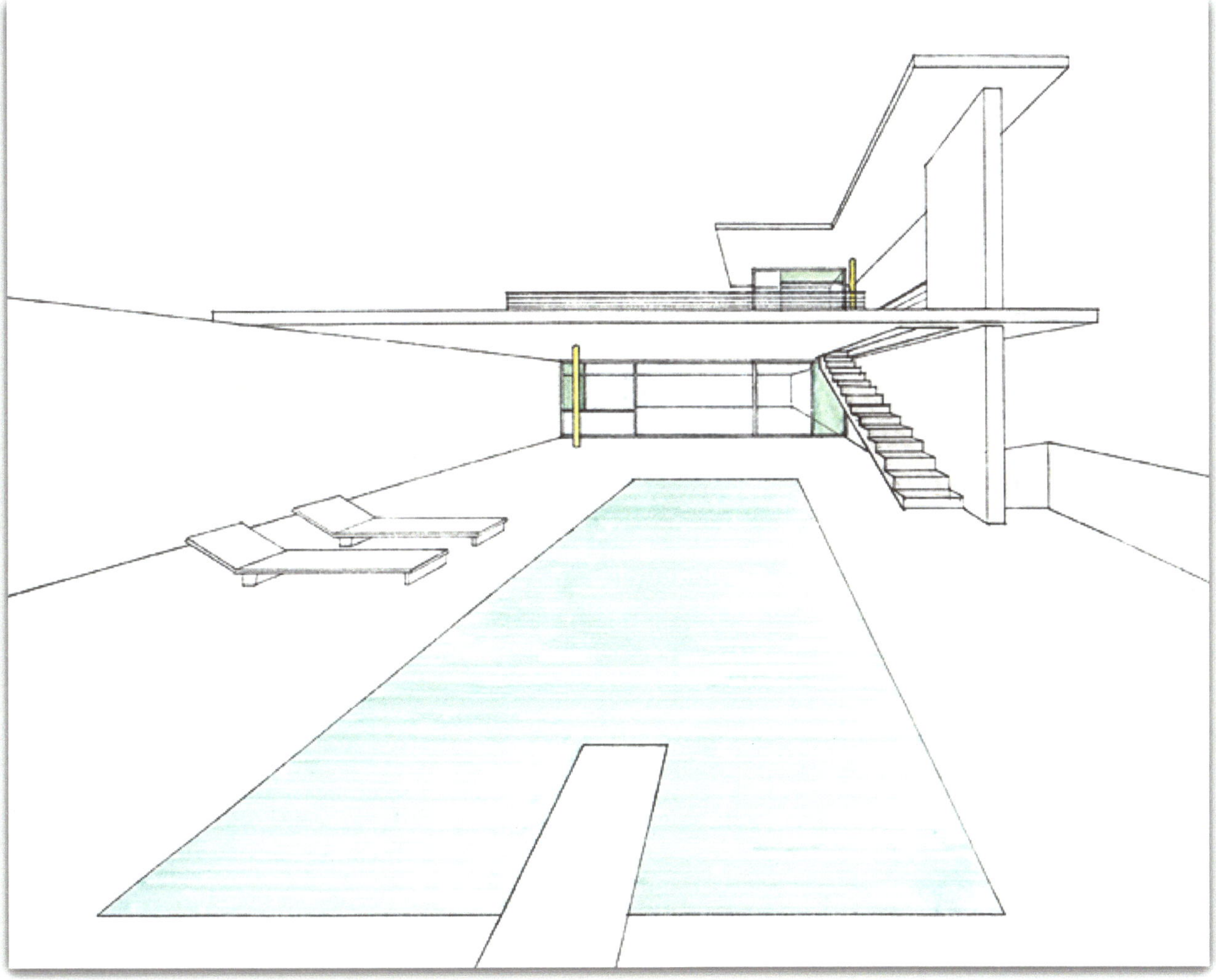

SEEING THE WORLD WITH THE OTHER "I"

Venkatesan
Subbunarayanan

TABLE OF TOPICS

TEST OF TEMPERAMENT

It would only be a mooted point to promise the reader that, anything would change in one's life to a great deal of extent after reading this book in certainty. Continuing from a never existing previous version of "Mass aN 'd Choice", here in called "The MC", the best that may result at the end of reading this complete book of 100 pages, must be a slight increase in the vocabulary of English Language, improved speed in reading skills of an avid, voracious reader and a possible phlegmatic transformation towards a better temperament in fighting monotony.

That being told the inspiration, of "I" writing a book with fictitious tales and real life stories derived from experimental and experiential knowledge gathered over 14 years, has its roots deeply attached with many of the authors, who have spent days and nights in putting a myriad of words together to gather the reader's attention for an hour or two or may be more in sporadic intervals. So, here with both "I" and the "The MC" can only thank the readers who are spending their valuable time, and possibly an affordable cost to buy the book and encourage a blissful voyage towards reading the rest of the chapters.

$$? \, Ƕ_2 + O_2 \; ----> 2Ƕ_2O$$

Amongst the topics of interest which are intriguing to me, I did want to start with a topic that I presume is a natural choice for most of us. "Intention" and "Temperament", two strong words to start a topic in this book which still has 98 pages to skim through to move on in one's life. Though phonetically, these two words may not have a great connection to each other, they do have an epsitemic value. They may straight be compared to the variables in a physics principle which states that "Momentum" and "Position" cannot be related and measured at very high levels of accuracy. In fact, the greater the knowledge of the first, the lesser the known fact of the second. For instance, the "Intention" of buying a book with the greatest aim of inventing a novel idea may just be against the "Temperament" of researching vast subjects and putting together a proposal with facts and proofs.

"Intention" on one hand provides the necessary platform to build a dream. "Temperament" on the other, gives a vivid picture on the readiness of an individual to understand the design slowly, gather the raw materials completely, create a cohesive bond strongly and test the product thoroughly. The slower the pace of "Temperament", the larger the probability of reaching the "Intention". Although, having only one of them in life may take one nowhere. It may be too early for me to set a context for this book, talking about an inspiring journey of an example's life till date, providing real world examples of how all the achievements created a "Feel Good" factor and helping transform lives around for a better living. Rather, the initial few topics of interest may revolve around the theories of interesting subjects which contributed towards a paradigm shift in genome of the mankind.

"Trying" is a great word which may fit in between the two words with which we started this book. We all know what "trying" means. However, few of us may not know what it results in. So, here I try to begin the book and highlight the importance of "Trying".

$$1\,\bar{H}_2 + O_2 \;-----> \;?\bar{H}_2O_?$$

The Former Concept of momentum and position may state that the initial push towards any object of "Intention" is all that's needed to start the goal of building a dream of "Temperament". Fascinating it may be, the level of momentum needed to create the initial force to gradually push oneself towards the "Intention", is almost always a multiple of momenta. Though, the physics principle states that the momentum P at a point of time is a multiple of mass (m) and velocity V (vector), when there is a gathering of multiple momenta, the direction remains to be considered an important tool for measuring the momentum and there by the position and intention. Let's analyze this with an example in crude terms.

"Anything and everything on the earth may be considered a mass". If a thing B has a mass of 6.64 Kg, keeping gravity as constant, it would result in an exertion of approximately 65 Newtons of weight over the earth. Weight varies as the consideration of gravity during motion is not accurate since we are "Trying" to measure weight when the mass is stationary. This also means that the mass, momentum and position are inter related as per the same principle of physics, since the earth as a mass by itself has a rotational speed.

Let's step back for a moment, and get to the sub topic of "Temperament". This word in theoretical physics would simply imply that, the thing B described in the passage which had a mass of 6.64 Kg, is now adhering to specific set of characteristics shared by several other masses. It may fall under any number of classifications, exhibit enormous similarities, show very meager differences, but when measured under the constant force of earth's gravity should justify and align towards the direction and magnitude of the momenta. In other words, the characteristic of mass B has a specific alignment towards Momenta M, with varying factor of Vector V.

$$? Ђ_2 + ?O_2 ----> Ђ_2O_?$$

If the "Intention" is set to a constant say "I", Temperament is set to a variable say 't', and we derive an equation out of Intention and Temperament, it may just adhere to the same principle of Position and Momentum described in theoretical physics. In case the momenta of earth's rotation is considered, this would vary to a great deal of extent since the variable which had to be considered as a constant is now considered a variable. In theory, if we try to get to a point of trial and error to balance an equation of two English variables, with an elimination of variables which are trivial and remove the by products of the reaction, then the equation must stick to the theory of stability.

Questioning, the authenticity of the two words in English, named "Intention" and "Temperament" by applying a theoretical physics principle and trying to analyze a confident step towards one's ambition by setting a relation between the two may slightly be demented. Let's assume, "I" as an author or a proof reader, miss to add quotes to the words which I am trying to analyze in the passages, the entire reading exercise till now may take confusing turns. Reason that can be attributed to this may be, that the words which had to be treated to be belonging to a language, were getting treated as entities.

An interesting question which pops up, "Do words really communicate with each other?". If so, how. Probability in Mathematics can answer this question. In my quest towards naturally occurring scientific phenomena, I forayed a little into the nature of words that used to trigger a reaction in the reader's mind through some of the best selling books and essays. Let's get to the topic of momentum and see if it can be used as an attribute in some of the theorems in probability.

$$? \hbar_2 + 2O_2 ----> ? \hbar_2 O_?$$

If the momentum of the most if not, all of the words in a 100 paged book, set 's a uni-directional tone towards a GOAL of achieving a position, then unless the words chosen in the book by the author communicate with each other in an adhesive way it may not be possible to substantiate the claim. Many readers may wonder why there needs to be a proper momentum set, to get the "Intention" in motion. Any mass in the universe has a chemical composition. Broadly, the composition can be classified as made of either pro-active or re-active elements. There is also a third classification named inert elements. If we consider some of the elements in the human body as highly reactive, some elements as pro-active and few elements as inert then, we can derive a possible connection between the elements and the effect that "Intention" may have towards both physiological and psychological well being.

A SPORTSPERSON, MAY MAINTAIN AN ATHLETIC PHYSIQUE OVER A PERIOD OF TIME TO REACH THE POSITION OF WINNING MEDALS FOR A COUNTRY. THIS NEED NOT BE TRUE FOR A COMPUTER OPERATOR WHO HAS A JOB OF DATA MODELING OR FOR A WRITER WHO HAS A NEED TO BE RESOLUTE AND CREATIVE IN HER OR HIS WORKS. Circling back a little, if the Intention of a thing B described is to move mountains, then thing B has to have approximately an infinite mass. Alternatively, If the Intention of thing B is to invent an idea which could possibly move mountains, then thing B has to gather a vast experience to put together the facts and still may prove to have an infinite mass not physically but rather through words and works. The second alternative can only be prudent, since there is no mammoth structural design proposed, rather a subtle force getting added throughout the differential positions, and finally resulting in a thorough transformation. Differential calculus relates to the study of the rate at which an entity changes forms. This has a wide scope not only in physics principles, but in various other hypothesis proposed by inventors till date.

$$? \hbar_2 + O_2 -----> \hbar_2 O$$

Let's assume that the magnitude and direction, in moving towards one's desired Position, are considered as 2 connected variables that can contribute to the required force and acceleration, then these two have to be measured accurately at every differential position so that the steps in the calculus are sufficiently but not infinitely small. A good example to illustrate would be a "Snail Mail" which existed during 1900s. In crude terms, `the snail mail used to deliver messages via letters in a very slow pace.` This means that there may not be any guaranteed delivery date and `the person who is receiving the letter had to build a great deal of a patience to read through the messages`, whether the letter holds facts of secret nature or a work of an open domain with effective romance added to it.

In other words, Differential calculus may simply be stated as trajectory of a mass. It finds it's use in many fields, one of them being physics. The magnitude and direction in moving the mass towards a particular position can be plotted on a graph with X axis as magnitude and Y axis as direction and the Point getting plotted being the mass M. Though this would right now remain as a two dimensional plot which would tell us an intersecting point at which the mass M, may be stable and stay on the space of paper, with modern day probability theories and charting methods, these can be projected as three dimensional objects interacting with each other at different positions.

Thinking of the two english variables "Intention" and "Temperament" without physics and maths, would just be the same as thinking of a skyscraper without an elevator. The provision of statistical tools and modeling graphs add a huge value in getting to the right mix of Magnitude and Direction.

$$? \, \hbar_2 + O_2 \, ----> \, \hbar_2 O$$

Plotting of two of the variables in a two dimensional space, using a mathematical theory to provide high levels of accuracy to measure one's position at this point towards his or her goal, would not only turn this piece of book or electronic device into a computer with injected co-ordinates but also, would go in the other direction of the principle of uncertainty with which we started this book at the first point.

Before setting a goal in one's life, the veracity of the goal can be analyzed in detail. "Goal" can be set to a nominal Position to work towards in one's life to start with. Let's analyze why anyone would need a goal to work towards or we can step further to pose a question on why anyone needs to work in the first place. It would take us to the initial passages of the book where we used to describe the two words, Momentum and Position. "Work" simply means that a sustained amount of energy getting dissipated towards an activity. Energy as per physics can present itself in various forms, some of them include chemical, thermal and sound etc... Energy dissipation itself needs initial momentum and may or may not need the magnitude and direction. In other words, the molecules inside a body react with each other to provide a force in order to achieve a mechanical movement to complete an assigned work.

The assigned work, could necessitate a pro-active reaction of molecules in the nervous system and elicit one of the forms of energy from the body such as thermal and sound. To illustrate with an example, the nervous system in the body is a subtle construction of an enormous number of neurons and axons. Two neurons in the body separated by a small distance provides the body a platform to create an impulse to make the systems understand that a work assigned needs a thermal, acoustic or mechanical energy to complete the work.

$$1\hbar_2 + O_2 -----> \hbar_2O$$

If the work assigned is from another system, or a body, it would still need a level of impulse through a specific medium to make the other system or body to understand that a coordination of the systems inside the body is needed to complete the dissipation of a specified amount of energy. Now, its crucial to understand the levels of energy needed to make a single system of body versus a large number of systems to complete the work.

To avoid making things complex, once we start measuring the levels of energy to make a system or a group of systems work, we will have to first measure the number of systems needed, the pro-active or re-active nature of the elements in the systems and the by-products of the process once the work is complete. Talking about the numbers, if the work to be completed is of humongous nature then the number of systems needed to complete the work would mostly be an equivalent to the integer multiple of the derivative of energy needed for completion of the work. This can be reduced through a simple way. Identifying the direction and magnitude needed for the work completion, and aligning the right mixture of a set of systems and allowing a reaction in a sustained manner if the outcome is known.

This may also be done through the traditional methods of trial and error, permutation and combination. In many systems or a group of bodies, there may be procedural lapses during this process, which may result in anomalies since we are treating work as nothing but a controlled chemical reaction. During the initial invention stage, many of the inventors may have adopted to the method of permutation and combination simply to understand the nature of elements in a minuscule environment.

$$2\hbar_2 + O_2 - - - - - > \hbar_2O$$

This would for sure need a manual intervention instead of an automated procedure, since an automated procedure would keep "Trying" until the equation is balanced through permutation and combination and may not worry about the reaction and by products of the various equations to be tried and eliminated to achieve the final stability. Some times, it may extremely be tough to balance the specified equation, since the number of input variables may not adhere to the principle of physics or mathematics and may be just be a production of a random number. Two things to understand clearly, one was the nature of reaction of elements which were unknown to the inventors. For instance, the inventors may be in a gray area to clearly make an informed decision to balance an equation until they know about the output elements and by products. And Two, the number of input elements that was unknown and hence had to be kept at a very minimal number by these inventors, thereby reduce the levels of energy needed to create a reaction. If in case the reaction is beyond knowledge, then they had to begin and research to understand the elements on both the sides of the equation.

If "**Goal**", for instance is set to a position which cannot be achieved in one's life time, then it would be wise to rework and understand if the Goal can be underplayed to a less dangerous level to avoid any repercussions. Provision of various elements to achieve the goal, would be the key towards the first step rather than the setting of a goal by itself. This also means that, unless one understands what is right and what is wrong in one's life, setting of a goal may not be a great idea to start with. Position and Intention, Momentum and Temperament, needs to be well understood before commencing the activity or work towards one's Goal. **Once the goal is understood, the body has to accommodate the goal over a period of time. This can simply be compared to the digestion of the food intake on a daily basis**. Unless the mind and the body is ready to digest a goal, it becomes tough to produce results.

$$2\,\hbar_2 + O_2 \quad ----\rightarrow 1\,\hbar_2O$$

The Intention of initiating this hand book with an inspirational word of "Goal", may slightly lean towards a refreshing model of attracting the reader 's attention. However, I did think this may carry the reader to a Position, where he or she can start thinking to embark on a road less taken and request assistance if not work as a stand alone entity to achieve a formidable Position. Many may wonder, if they deserve to achieve or can derive the necessary energy to complete the objective. Answer may lie just in the ability to move to next level from where he or she presently is and then rework on the goal incrementally to avoid exhaustion. Provision of the necessary support needed, theoretically and historically has proven to be a cake walk, with a large number of elements providing incremental and reinforcing acceleration, without being asked for, in case the direction is positive and magnitude is minimal.

One may just be little cautious in deriving the energy and support needed to achieve his or her goal, since an innumerable number of forces helping to produce a great work may just fall under an artistic classification and skewing towards a work of public domain. Possible innovation in one's own work, protection of the end result and step by step incremental analysis of intermediate results towards production of a tangible model may prove to be a worthy cause to measure the lifecycle of a goal by itself.

$$2\,\hbar_2 + O_2 \; ----> 2\,\hbar_2O$$

THE 0 FACTORIAL

Infinite Loop, they say is the cycle of birth and death. The word '**ZERO**' has a prominent relation with the Goal one may try to set in one's life. Imagine a computer box getting fed with a recursive input of 'n' not being an integer and not 'zero'. The result will be an infinite loop and machine overload. The mainframe and traditional systems during the 1900's used to give a continuous beep indicating their disinclination towards this operation. There may be various ways to produce an infinite loop and there is also a minute difference between an infinite loop and 'Not a Number'. Any number other than zero divided by "Zero" results in 'Not a Number', multiplied results in 'Zero', added to and subtracted from results in 'Same' number, positive or negative respectively. Factorial is one of the subjects of interest, that can be applied in goal setting. If the question is how, Answer is Probability. We all know that the probability is the chance of any event to happen at a point of time in a thing B's life. Factorial may simply be stated as a set of all elements that are encompassed for measuring probability of an event to occur.

If the goal to be set is discontinuous at various points in one's life, then it is important to measure the probability of success and failure at these discrete points, help reinforce at each point by providing expert guidance. The denominator to calculate

Recursion (n) = R *   Recursion (n-1) = (R$^{\#}$) *   Recursion (n-2)..

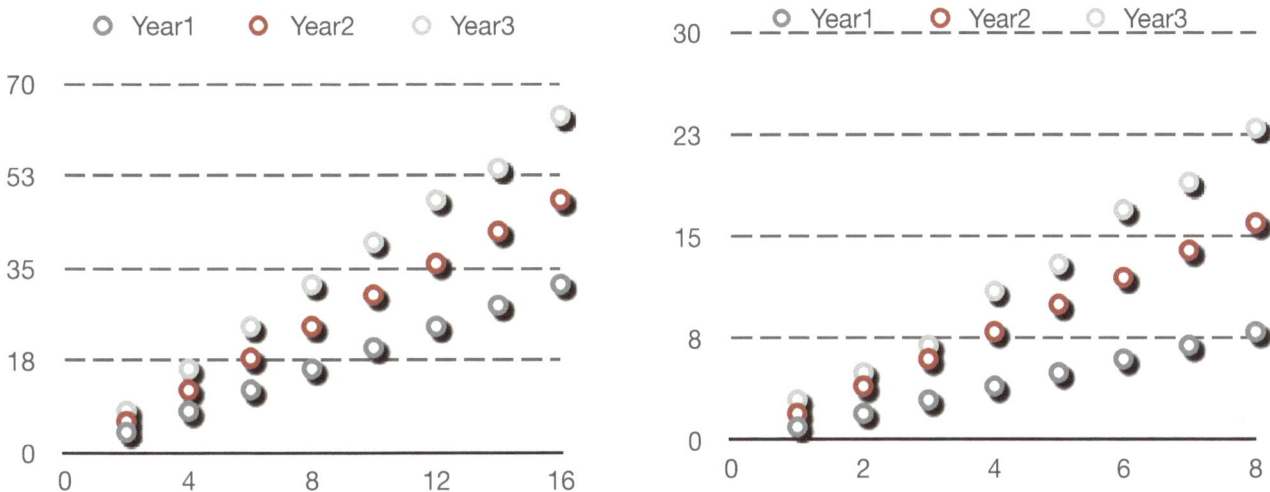

the probability of reaching one's goal, would be the distinct factorial of the events that already occurred towards the achievement of the (N-1)th position considering N as a goal.   The numerator can be the factorial of summation of magnitude and direction at distinct different points in descending order. The resulting fraction may just provide us a direction towards the percentage of efforts needed at different positions, momentum applied versus the momentum needed. This can be plotted on a graph with chosen co-ordinates to see if the plot is gathering right momentum, needs additional input, losing traction or out of track altogether.

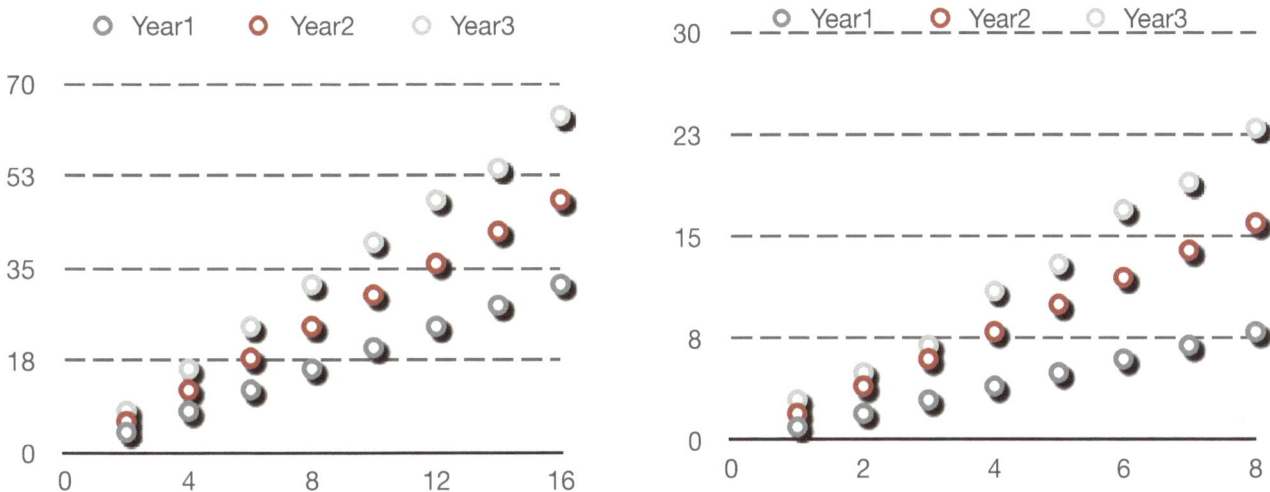

The above two charts illustrate the differential points at which the coordinates are plotted with energy dissipated as a multiple of 2,3,4 in the first graph   and natural numbers, even numbers and prime numbers in the second to achieve a limited sample set. Clearly, the magnitude of energy spent to achieve the (N-1)th position in the first graph is greater than the magnitude of energy spent in the second, considering we are in year4 of measurement. The first graph may be associated to a limited chemical reaction to achieve the work intended with an integer multiple N acting as an acceleration unit. Another factor to observe may be that the space between first and second compared to second and third line of plot varies to a normal range in the first. The motive behind considering natural numbers, even and prime numbers in the second graph is to analyze the time and distance correlation of these points so that a feasible solution can be derived on when, where and how a goal can be achieved, what can be the level of energy to be spent and at what intervals.

Recursion (10) = R *   Recursion (10-1) = (9) *   Recursion (10-2)..

Analyzing the graph 2 further, on the lines of time and distance to achieve one's goal, we can see that the distance between the $(N-1)^{th}$ points for the two line plots of natural number and even number, the distance travelled by the plot is larger for the second and hence reaches the destination sooner with double the magnitude. If we consider this during the third round, if the line plot is not provided the needed force, as per the pendulum theory of physics the line plot may slightly shift lower and reach a lesser distance and the distance may also slowly reduce in magnitude. Over a period of time and number of incurred frequencies, the line plot may stop traveling since it is presumed that the energy is derived from an external source and the external source may lose energy slowly as it dissipates to 'Y' number of other plots. This may ideally transform the line plots to a convergence towards infinity as shown below.

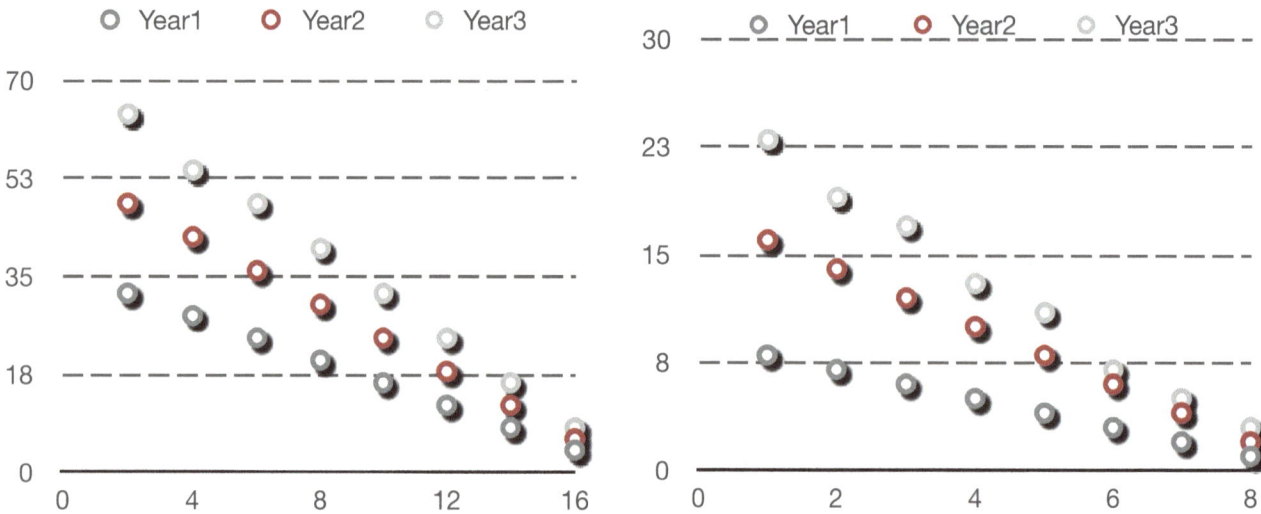

'ZERO' a number for sure, has a good significance in mathematics. LIKE, ZERO FACTORIAL AND ONE FACTORIAL BOTH RESULTING IN ONE, WHICH INDICATES THAT THE POSSIBLE ARRANGEMENT OF ZERO WITH ANYTHING LESS THAN ZERO IS ONE AND ARRANGEMENT OF ONE WITH ANYTHING LESS THAN ONE IS ONE. One may get perturbed on why a zero factorial and one factorial results in the same result, since both are considered numbers and a one factorial should possibly result in two arrangements such as 01, and 10 if considered in digits form. The problem statement here as per the perception of the reader would be to consider numbers in digit form, make an arrangement of these numbers in a paper and add value to zero, then the factorial values may slightly be higher in their end results, with 1 factorial resulting in two arrangements and two factorial resulting in 6 arrangements. However, this logic will not be true for applied mathematics since 0 and 1 are considered as arrangement of numbers and not as digits. So, a computer system and language considering numbers in digits form may slightly vary in its behavior when compared to human brains which adhere to rational decisions using the perception and knowledge base.

Let's presume that a mind reader, as a human evolved to understand the behavior of other humans well in advance than the event to happen, then he or she has to ideally read through a large number of predictive behaviors and make few deducible combinations and get down to a theory of elimination and finally present a set of facts which must mostly be accurate. In my quest towards probability theory and to understand the nature of this mystery, I researched further with a considerable event at hand as trivial as expecting same heads out of two independent events, when two different coins of different sizes, denominations and masses are thrown at the same time. Per the probability theory, my expectation of getting two heads at the same time would be 1/4th of the total number of events that may have occurred by throwing the two coins. So, increasing the number of trials to 4, Ideally I should have seen two heads occurring in one of four events every time. It did happen the first time, but not during the second set of trials. Reason may be attributed to various causes, but the probability theory clearly states that during a single set of trial, the chances of getting two heads at the same time is 25%, which is 1 out of 4 trials, however the chances do increase with multiple attempts in a larger sample set instead of a single set of 4 trials.

Goal, as a precedence in one's life can be projected as a treasure in one's daily routine. When one is able to assimilate his or her goal over and over again in a positive direction, it finds it's route through a spiritual channel. We can explain this through factorial combinations. If there is a failure in moving one self towards the $N-1^{th}$ position, the body's reflexes start auto adjusting to reach the position and if the energy spent physically and psychologically is measured it may follow a path of a factorial, since on a day to day basis the goal has been assimilated in the nerves through the sensory organs which are nothing but a

Recursion (4-1) = R *   Recursion (10-8) = (R#) *   Recursion (1-0)..

group of electrical connections. If there is an exertion of a large force by the body to move towards the $N-1^{th}$ position thereby resulting in a great deprivation of energy, it exerts a greater mass and hence a larger movement as shown in graph 1. Though the position can be reached quicker, it also results in exhaustion and hence $N-1^{th}$ to N^{th} position movement becomes difficult to predict. It directly compares to the example of Hare and Tortoise race, provided both are healthy and of the same mass. Movement of Hare equivalent to the deprivation of energy with faster sprints, movement of Tortoise through infinitesimal steps.

Though the Tortoise may win the race in the traditional stories, Hare may have just provided the initial momentum for the tortoise to move towards the destination.

In a Zero factorial theory, there is no winner or loser, rather the exertion of energy is measured at different points over a period of time.

During the observance, either the Hare or Tortoise may be chosen to progress towards the designated path at different speeds.

The e 'nd goal, may just align towards the adjustment of the right energy and right magnitude with the initial momentum and needed protection.

Many a times, Goal works as a protective force, in one's life. Since the Goal sets a conditioned path to compete in a designated track like a track assigned for each swimmer in an olympic race. If the goal set is a common one such as, gardening which may also be treated as a hobby, then the time spent towards this activity will be minimal, momentum be small and direction be positive. One can try to observe, if this gets converted to a successful activity by performing the activity consistently for at least the minimal number of trials mentioned in the graphs. With a good temperament, one may taste success which would instill the needed confidence to continue the activity in the designated track with the other set of activities getting sidelined. **The proof of concept would reflect from the outcome on a daily basis.** Many inventors and recent day scientists, who had struggled to uncover the mysteries in understanding the human potential were able to manage to achieve their respective goals by investing time and energy to study and understand many of the equations thoroughly and revealing secrets in their fields to help larger mankind, be it a medicine or a locomotive.

Recursion (11-6) = R *   Recursion (9-4) = (R#) *   Recursion (5-0)..

Goal, can be analyzed on an infinitesimal space in the graphs with magnitude and momentum getting plotted with reduced intervals. If the graph which we used previously, has a limited number of trials and improved magnitude and momentum, it would result in a continuous range and show us a vivid image of the position of the goal setter and let us know where we are in the space. In case the space between two trials are set at distant intervals, there is always a risk of one losing the momentum between the trials and hence losing magnitude. It may also result in the curve taking a different direction if the momentum is varying at different intervals. Let's try manipulating the graph plots with 2 of the input sets kept as same as previous graphs and one of the sets reversed in order to understand the position at which this intersects and gives us a position where a possible synergy happens.

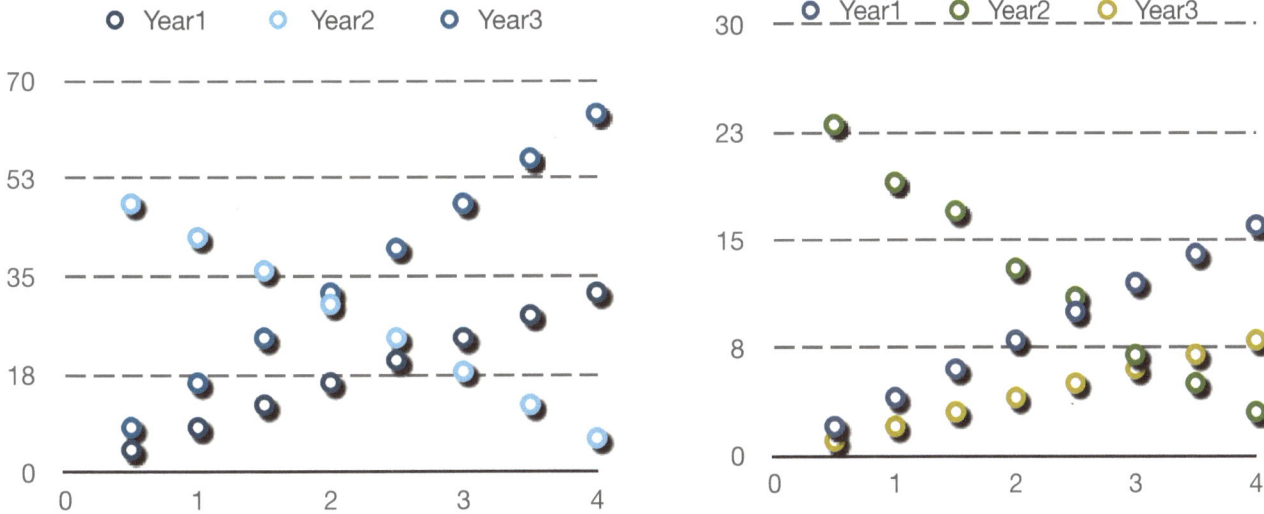

Clearly in graph 1, the intersection is at a point above 2 in the x axis and in graph 2, the intersection is at two points, one at 2.5 and another at 3. Remember, the graph 1 was plotted with y-axis as multiples of 2 with higher magnitude and graph 2 was plotted with y-axis as natural numbers, even numbers and prime numbers with lower magnitude. Though one may see the intersections happening at a later space of x-axis in graph 2, one needs to understand that the magnitude at which the intersection happens in graph 1 is around 33 and in graph 2 is around 13 respectively. This will imply that the position that a goal setter may reach with a greater magnitude will be around 33 in graph 1 and 13 in

Recursion (6–5) = R *   Recursion (7-6) = (R#) *   Recursion (8-7)..

graph 2 with 2 and 2.5 units of exertion in x-axis space with x-axis being treated as time, mass or any other unit of physics which we discussed in the passages.

This can be compared to microscopic vision versus bird's eye view.

In a microscopic vision, all that we see are the particles at the minutest level. Be it a living cell, atom or an electron. When a mass needs to be accelerated, the particles inside the mass have to be excited to the level of exertion of force needed to accelerate the mass.

In a bird's eye view, all that we see are the connections of the particles, of the living cells, and of the atoms or electrons. The higher the bird's eye view, the greater the illusion of the stillness of the connected particles and a sense of satisfaction towards the achievement of the goal.

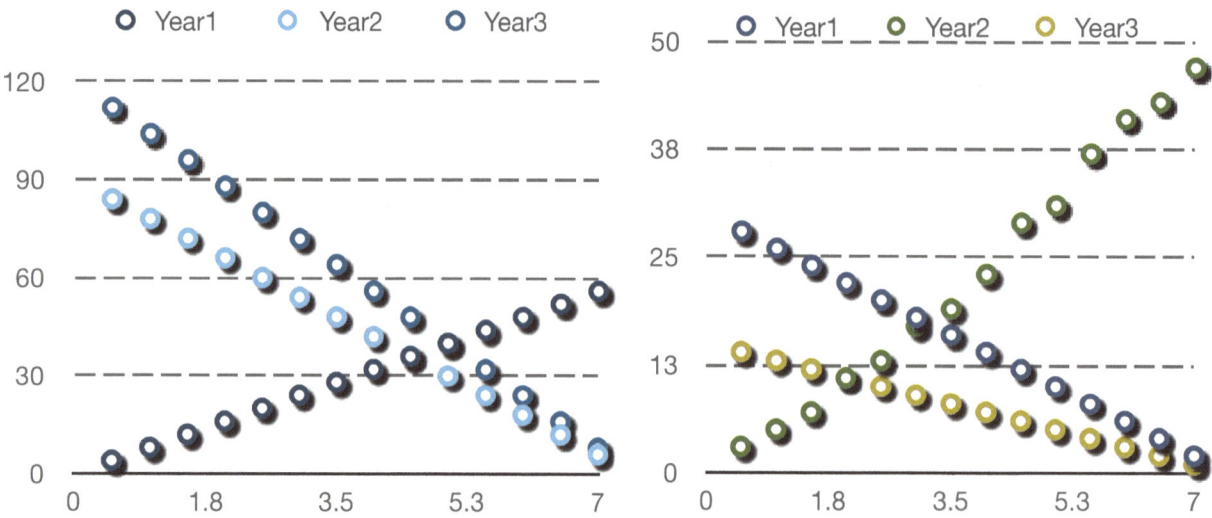

Let's even step up slightly to reduce the intervals between the x-axis units and improve numbers in y-axis to provide a clearer picture of intersection. The x-axis would now have 6 more additional units and y-axis will have respective multiples of 2,4 and 8 in graph 1 and natural, even and prime numbers in graph 2. We can notice that there is a

Recursion (7–0) = R *   Recursion (8-1) = (R#) *   Recursion (9-2)..

perfect synergy at a point approximately around (4,32) in graph 1 and a 75 percentage synergy at around (3,15) in graph 2. The graphs go on to explain a clear principle that when the plots are increased in number, there are high chances of a complete synergy and energy production through a reaction when compared to minimal number of plots shown in the previous graphs. They also go on to show that there is a better probability of intersection of these masses in multiples of natural numbers than in prime, disjoint sets. It converges to our original theory on the angle of inclination (in other words direction), magnitude and momentum towards the position to be achieved. Let's assume if the graphs are made to show disjoint sets of numbers such as prime numbers and even numbers as shown in graph 2, there may not be intersection at any point if the direction is the same. An attempt to disprove the theory is to change the angle of inclination until there is an intersection which would mostly be an ascending versus descending order of the two sets of numbers.

The original theory of uncertainty principle discussed in the initial portion of this book, in which the rotational momentum always peeps in anything and everything we do as part of every day attempts towards the goal. The tangential plot shown in graph 1 and the angular plot shown in graph 2 provides a great illustration to be compared to the theory. Let's understand the point that the angular plot shown in the graph 2 holds a valid premise against the linear plot in graph 1 because of the sheer wear and tear of the equipment, tool or a mass which we use to reach our goal. This can be compared with the construction story which most of us already know about the time spent to sharpen the tools used for construction versus the construction time. The larger the time spent to sharpen the tools, higher the precision and lesser the time in the outcome of construction.

The angular momentum holds an interesting key towards the achievement of one's goal whether its a short term or a long term one. Let's say, if the momentum is linear and we start working towards the goal relentlessly, by inputting the extra energy in a linear fashion, it will result in an exponential plot and there needs to be a reversal of the order in case of multiple plots to achieve synergy. However, if the plot is angular and momentum is adjusted towards rotational or slightly angular, then the original plot has to adjust itself towards the other plot and intersect to provide synergy. Angular momentum finds

Recursion (2–0) = R *   Recursion (3-2) = (R#) *   Recursion (8-8)..

its applications in many industrial needs including locomotives, piston pumps and things we use on a day to day basis such as fans and cycles.

One may even go to the point of questioning why anyone needs to consider a planetary position before commencing an activity on a day to day basis which may help move them towards their goals. This proves to be a vital point to consider since, a large number of masses working towards a singular goal of achieving in their fields, be it medicine, technology, law or accounts for example, the number of masses in each field have to be balanced in order to reduce or increase the energy levels needed in the respective fields. If a particular field of study or sports has a large number of masses working towards the goal, then the energy spent towards the field increases to a great proportion and hence it becomes a linear plot instead of an angular plot. So, it becomes multiples of integers as shown in graph 1 and not a list of naturally occurring numbers which would have angular momenta associated with them. The list of multiples shown can be compared to list of fields in which one has chosen to study, play or perform. The sample set mentioned here when assigned a weightage and made unique, may result in no intersection for a synergy at any point over a period of time measured in x-axis since the field may prove to be disjoint instead of naturally occurring numbers.

When all the planets are considered rationally in the solar system, and assuming only Earth provides favorable conditions for living beings till date, we may say with high levels of confidence that a great number of masses working together towards a singular goal in only a single field of study or sports may result in an exponential plot and a larger energy decay instead of an intersection towards a synergy. It may also cause inter-planetary angular momentum distortion since the greater number of masses working towards the goal will have an effect on the Earth's angular rotational field and hence a disturbance in the gravitational force which keeps all the planets in the orbit rotating till date. The evidence that many parts of the Earth are still facing extreme climatic conditions may just bolster this theory, that the Earth by itself is trying to balance the reducing energy levels caused due to varying angular momentum at different locations and carried towards the other regions with the help of particles and waves.

We, trying to prove a point in this session by furnishing an example of Hare and Tortoise race and plotting different masses on 2 graphs, one with multiples of numbers and another naturally occurring numbers to step towards one's goal may just provide us an instinct on how a goal needs to be set, which direction it needs to be aligned, how many masses may need to work on the goal to provide a strong foundation, how to adhere towards an infinitesimal set of steps on a day to day basis and how to provide the necessary momentum during catastrophic conditions. A plausible note one should take up on setting of a goal to work on, is to analyze the goal well in advance, dig deep to establish the facts, ascertain external environment to avoid "sit and wait" opportunities and use the natural and artificial momenta one may "see and feel" during the due course.

KNOCK, A REALITY CHECK

Augmented Reality, is a field of innovation and development, to the extent of locating the right vein in the body to make a perfect diagnosis. The field has made its mark till date in a gamut of industries encompassing Technology, Visual Arts, Education and Military. An interesting day to day application, is that it can show how you look on an apparel, with a stroke of a button and make you look better with another stroke. **To assuage you in moving towards your goal on a day to day basis, the concept of Augmented Reality has found its way to innovation's peak.** Imagine a world without smart phones, tablets, computer systems, locomotives without smart meters or even televisions without remote controls. There is a level of smartness getting added without anyone's knowledge in the air that we breathe with a by product of chemical reaction between the electricity inside our own nervous system and that of the one produced by the devices.

Quoting this with a simple example would be to compare two distinct events of a human predicting an event to happen in the other part of the world through tele-transmission and recording it using a magnetic tape or a compact disc in this century versus a human predicting an event to happen in the other part of the world using conceptual physics or applied mathematics in the previous centuries. Even before Light as a medium of usage was introduced by the innovators, the concept of Augmented Reality would have found its base in day to day application. For instance, most innovators would have kept sunlight as a time keeping reference to warn themselves on which part of the day they were in and what they had to do at that point of time. They also would have made use of their vision to locate the right veins, bones and other parts of the body during different timezones of the day to understand the effect that heat, light and moisture has on the body.

There is a subtle difference between Augmented Reality and Virtual Reality. 'Augmented' in English Literature means addition. Augmented Reality in general adds to what is existing with you already and tries to improve the user experience on a daily basis. Virtual Reality takes us to a different world altogether and provides us a virtual experience. For instance, when a human, who had been making a living on Earth for a very long time, lands in the moon for a brief period of time and comes back to Earth, the brief period could be called a Virtual Reality though it is actually a true event that happened. The same could be compared to an advance stage of a REM sleep in which one forgets about whats happening around and in a different world altogether. Virtual Reality may also be attributed to something which is technically impossible for a physical body to achieve, however the psychological force of a physical body can still achieve. **Augmented Reality on the other hand, adds provision to pacify the ability of the body to understand various aspects to step towards the goal in a non intrusive way.** In the field of education, in order to fight the boredom of reading the same story told in a different way through enormous number of chapters, the augmented reality can help understand the concepts faster, nicer and provide monitory support to see if there is a decline in the health because of exhaustion.

Think of a spectacle glass which tells you how much will be probability of a car approaching in the opposite direction, switching to the lane not specified for it to travel while you are driving your car in your lane. With increasing number of vehicles on road, an assistance to provide a far-sighted vision for a driver to foresee the circumstances would be a great value add, however an increased dependency towards the augmented reality will result in augmented reality becoming virtual reality by itself.

$$2\,\text{AgR} + 1\,\text{Vr} ----> 3\,\text{VR}$$

Augmented Reality, can be applied to our aim of goal setting, in which we try to see how innovation in a space of paper may trigger an idea in a reader's mind. Let's presume a goal setter has a set of 3 goals for now, and each of it is a short term goal. If each of the goal is provided the needed momentum (force) at specified intervals of the months such as April, May, June etc.. with momentum as random numbers and Position to be analyzed in the space of paper, we will see a plot shown below with an Augmented Reality plot. One may wonder how and why this will make a difference in a goal-setter's life and a goal's life cycle. We can clearly see that in a 3 dimensional space, there is no intersection or synergy needed, and each of the goals are provided independent care and attention. Also, it provides a detailed analysis of spatial interpolation of each goal, the ability of the goal to move ahead without interference.

The interpretation of the 3 dimensional plot is tricky, in which one may have to keenly note the position at which the goal setter's activity started. Though the x-axis and y-axis seem to be quite normal compared to our previous plots, the additional z-axis makes a difference here since the z-axis provides the necessary cushion for the goals to be achieved smoother with less collision. The same can be seen in the blue and green plots which are aligning with the angular plot and the brown which aligns with the linear plot. It's also interesting to note that the same piece of a device or a paper can be used to project 2 dimensional and 3 dimensional objects in a non intrusive way. **Augmented reality can simply compared to the addition of constructions, roads and buildings to an Earth without buildings, borders and only a separation between ocean and land.** What we perceive on a day to day basis as an object which we can touch and feel, adds to an Augmented Reality.

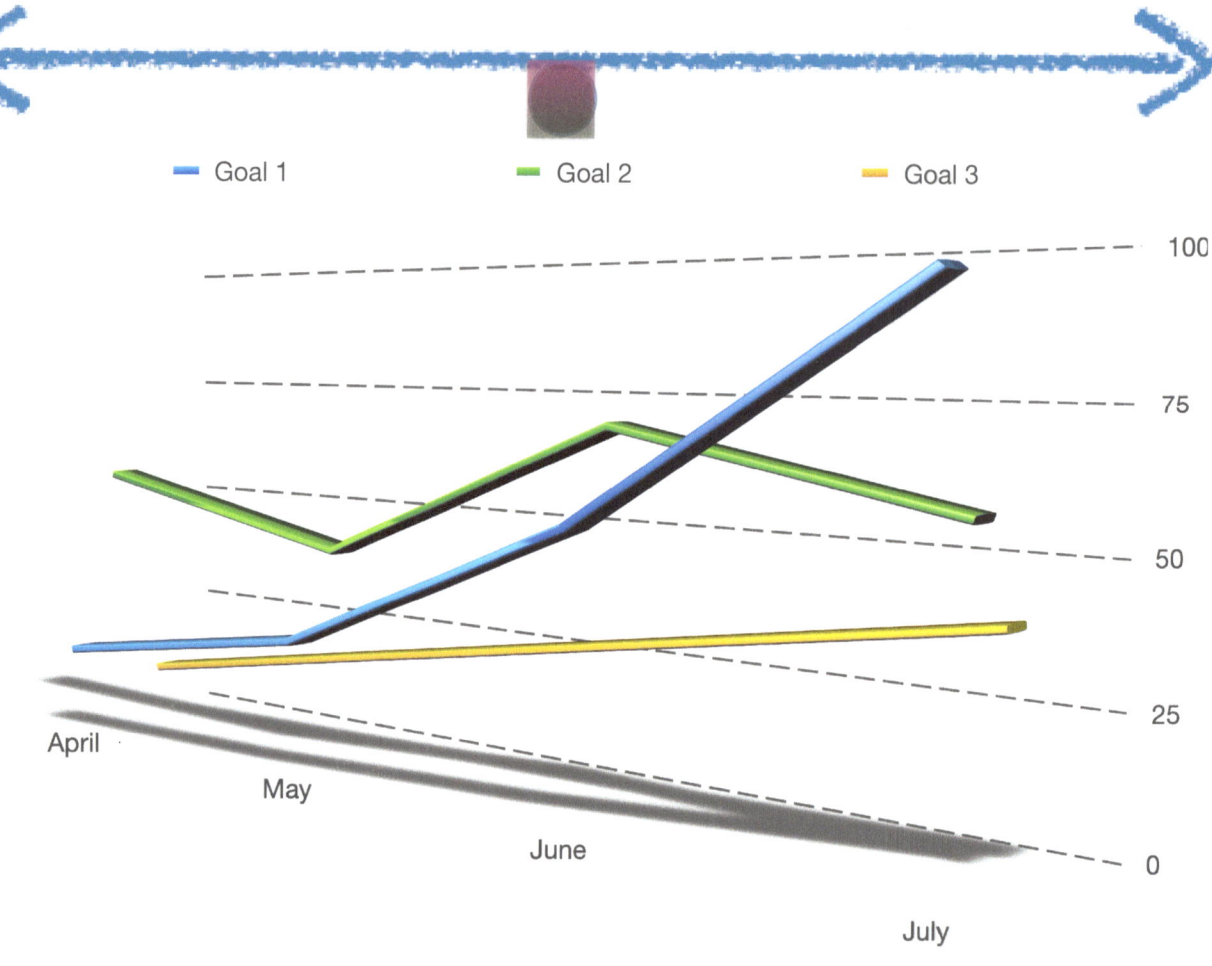

A simple example for an Augmented Reality Goal for a movie maker would be to provide a 7-Dimensional experience for the movie goers in which the movie goer gets to see, feel, smell, taste and hear the experience which she or he would get when visiting the location mentioned in the movie. Though this goal would be a long term option in the making of a movie, the short term goals would be to gather initial inputs on feasibility, understand if the entire exercise is worth producing for the movie goers. This would in fact circle back to the original idea of virtual reality in which we close the option for the movie goer to know if the actual portrayal mentioned in the movie is existing or not.

The idea of providing a spectacle to watch a movie with hidden facts displayed in the spectacle for the audience to solve may just be an added value for the movie maker.

The spectacle we are talking about to watch a movie would be the same to guide the audience to step out of the theatre and find directions to go home or inform about an emergency need to be attended to for a relative.

There are 3 dimensions in this aspect, first: the big screen using which the movie goer watches the movie, second: the spectacle with Intelligence which assists the movie goer to understand the movie and third: the movie goer's Perception.

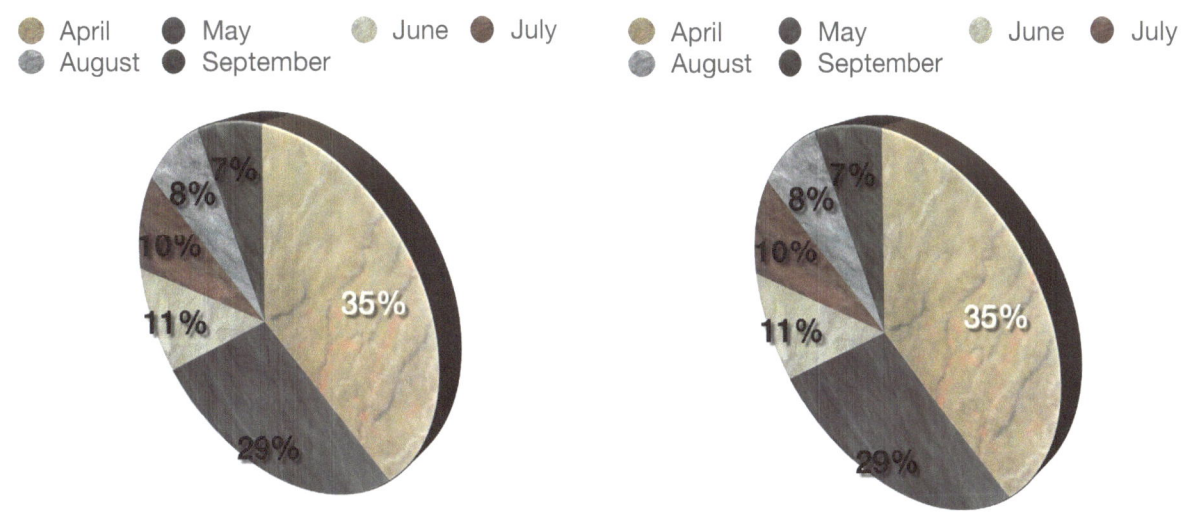

It may be luring to know about a spectacle which can assure us a clarity vision during old age with warranty, and help see a good world. **Spectacle** for me is a great innovation, which finds its way to protect us, help strengthen the vision by resting the mind and many of the sun glasses which control the radiation and reject extremities in the nature of the things that we may see accidentally. The above two pie charts can be compared to a 3 Dimensional spectacle for a person with irregular vision. Let's assume that the person with irregular vision is able to see only few objects and having a blurred side vision, then a 3 dimensional spectacle which can auto adjust the levels of light entering inside the nervous system for us to perceive the objects clearly would be a right medicine. The lesser the nervous system's ability to adapt to the objects that we see or hear, more the control we lose towards the Augmented Reality.

The above two charts portray a random assumption of usage of the vision on the right for a right hander who is programmed to perform most of the daily functions using his or her right limbic system. The nervous system has a natural tendency to continue the usage, since it has reached the comfort level already and tunes the body to perform anything and everything towards the right. And many of the laws in physics, have told us that a mass in the universe has to have a wear and tear. Here we consider our body and mind to be separate masses, since the body exerts Weight on the Earth and the mind exerts Pressure on the Atmosphere. This implies that most of the energy spent to the activity of seeing anything to complete the activity assigned towards the goal has a profound effect on the wear and tear of the nervous system that is used for perceiving the vision. Even if the goal-setter is conscious of the fact that she or he can spend minimal amount of energy on a daily basis to inch towards the goal, it adds to the training of the right limbic system to a great extent since the time taken to reach the goal in this case is long.

We can also see that there is a small portion of black region in the pie chart, which lies in the lower part of the figure. This implies that there is a drainage of part of the energy stored in the form of potential, kinetic or hydro electricity since Earth as a mass, always exerts gravity on the human body in upright position. This can be alleviated by providing angular momentum to the body or making the body stationary at different positions. If the Goal set by a person is highly ambitious, it takes equal amount of energy to move towards the vision. The remaining four quadrants shown in the pie chart goes to show about the other short term goals, which may or may not be related to the long term goal. The lesser percentage of energy spent also helps to cushion and rest the remaining parts of the body when one of the limbic system is active to a great extent. One can say with good confidence that the energy expenditure versus the energy gained during the entire activity will be the delta needed to sustain the activity at equal interval or continuously if maintained well.

The energy equation plays an important role in the maintenance of stamina in major long duration games such as tennis, basket ball etc..

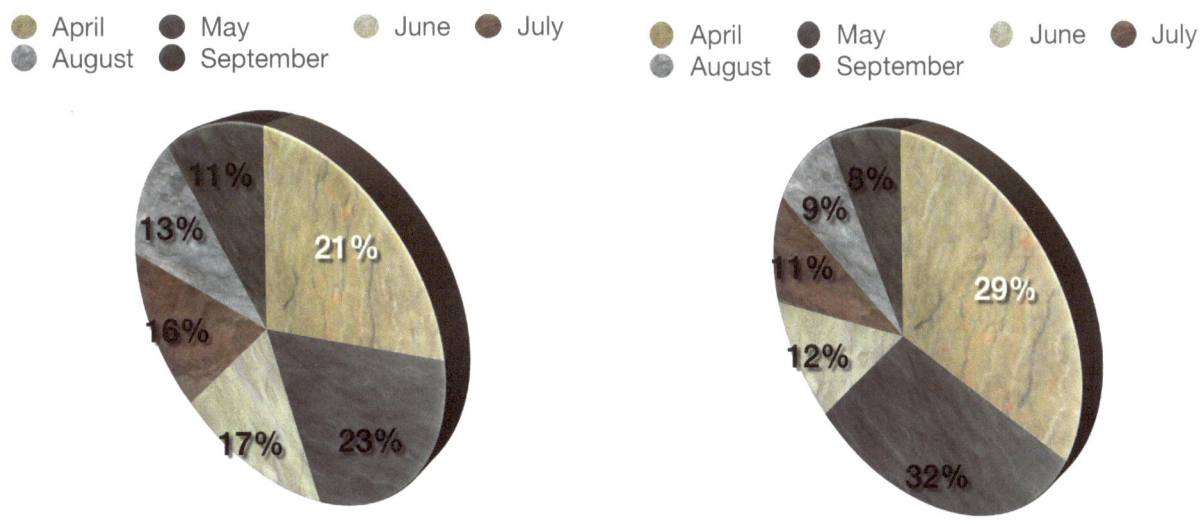

The above 2 pie charts are now tweaked to provide a slightly comfortable walk towards the goal instead of a great deal of strain. It is as simple as using the body's own nervous system to provide the needed energy to the dominant limbic system from the recessive limbic system. This means that the goal-setter knows how to use the nervous system for the perception, how to assimilate and sense any strain and provide rest if needed. When the limbic system is giving up, the goal-setter may have to take external support to take action and elicit energy from the other limbic system without affecting day to day activities. In sports, the player who is successful during the course of the game generally maintains a composed demeanor and preserves a great deal of energy for completion. An interesting point to know is that the nervous system keeps fighting with external invaders most if not all of the time in a day or night. Any Goal which we set and work towards, need not be enervating to such an extent that the duress gets added to the existing strain and reacts <u>with</u> the nervous system. Instead, the position to be reached can be made less far by calculating the level of energy needed and the level of input to be provided for both mind and the body. Input for the body may be nutritious food, simple exercise and input for the mind may be rest, creative thinking for less time and maintaining alpha waves.

Most goal setters in the field of sports know well in advance about the time needed to rest the body and mind to walk in the path designated. In sports which need a group co-ordination, like hockey or cricket the effort becomes slightly complex, since the mind and body of every player need to be considered. It will in fact reduce to a simple equation of <u>not</u> having every single player as a top performer since the field becomes a collision path for a smooth performance of the system to work towards the goal assigned for the team. It would make sense for a selection of trainee players, experienced technicians, captain cool and non-controversial leaders.

it may be ideal for a reader to go through a book of 150 pages in a single stretch and get an overall idea of what a book is all about and then start reading in detail to know about the nuances and the original intention of the author in each of the topics. This may not cause an exhaustion in the reader's nervous system since there is no need to remember the topics, reading will only be a part of the daily routine and not the only activity and enlightens the reader towards a better path. On the other hand, if reading is the only activity proving to be the bread and butter for a goal setter, then it does cause a tremendous strain if one has to use the nervous system to understand the concepts, analyze and produce results. The reader at this point has a choice either to read or ignore. Mostly in one's life, day to day work boils down to simple decisions that one makes and the causal effect created due to the decision. The decision making tree and algorithms used in the recent day machines make a strong foundation of premise in producing accurate results for a user.

In an Augmented environment, the machines which are augmented with the user's products do interact with other set of machines to produce a valid proposal for the user to adopt. **If a goal-setter believes in Augmented Reality and is able to make judicious calls at every point of time in her day to day life while using these equipments, then it would result in betterment of the environment** around the goal-setter provided the Augmented Equipment is able to adhere to the set of rules it is supposed to be working against, measured in time.

The below 2 charts reveal a choice for the user to perceive the same chart, same co-ordinates but through different angles. The first chart shows goal 1 leaning towards lower part of x-axis and slowly reaching the peak. The goal 2 starts at midway and ends at mid way. However, chart 2 shows the Goal 1, starting at the highest level and going lower towards zero in the x-axis curve. Goal 2, though reversed, maintains the same levels as in Goal 1. The reason being, the inclination of the chart 2 is made to only see the results in the beginning and not through out. **As many of the philosophers mentioned in their works, that a huge ocean is connected by tiniest forms of molecules of water with large levels of currents produced by air and resulting in waves.**

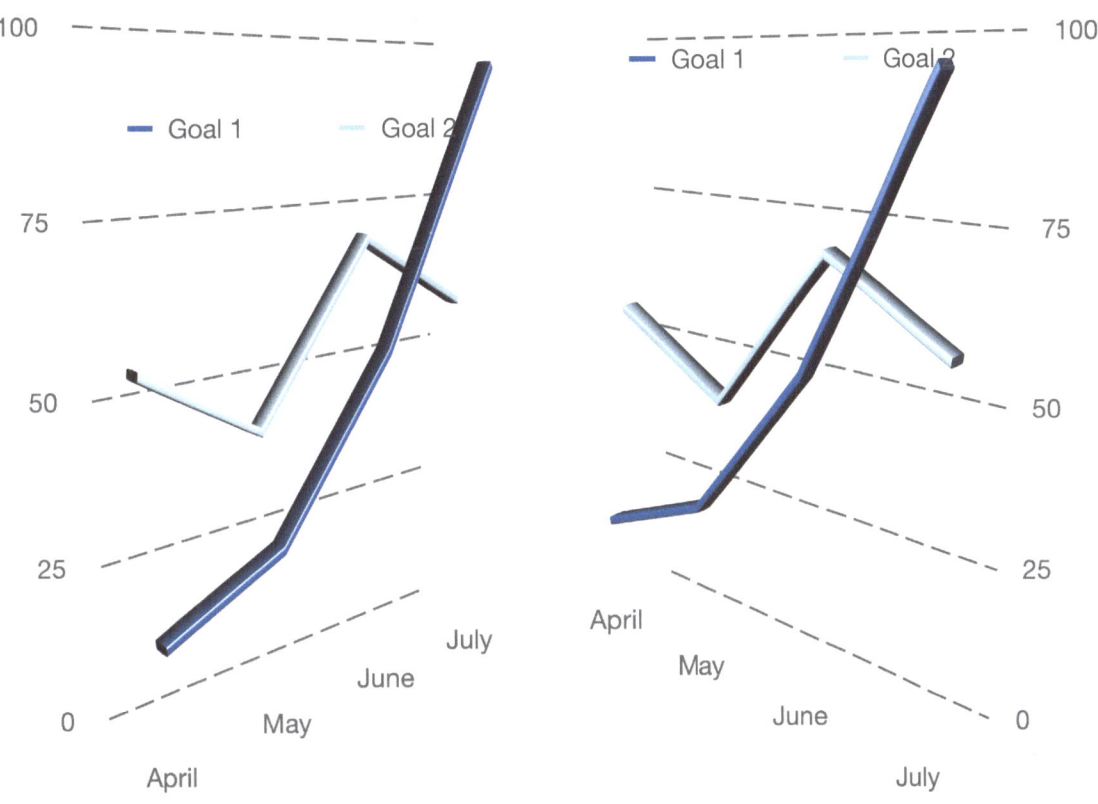

Unless, the formation of a goal is connected well by tiny molecules of water and becomes inseparable, the goal by itself cannot produce an electric current needed to run the day to day activities necessary to produce great work. When we see a great work or invention produced in the field of medicine, an elixir for life, we will have to go through the level of efforts, a healthy vision to move towards the vision and the path to be sticking to until the activity is complete.

The below set of pie charts connected with a thin line implies a goal-setter's ability to be connected to the day to day chores, maintain a physique needed to survive the present day challenging scenarios and present the position victoriously by persevering and adhering to the set of principles which may slightly differ from the goal of an Augmented Equipment, which would be able to produce instant results with an input logic.

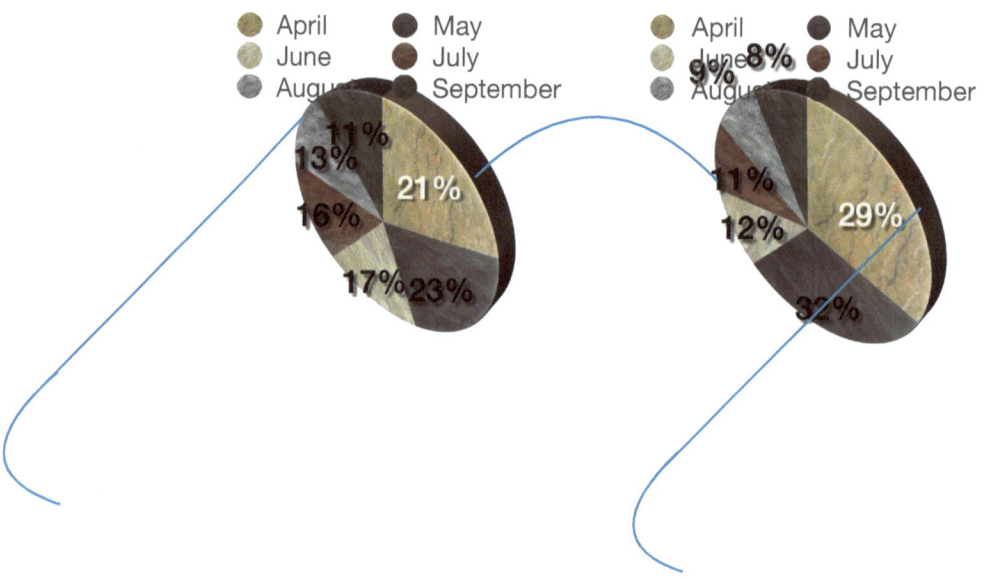

PAIN

Pain, a great topic to read and write in one's life. The innate mechanism of one's physical body to reduce pain slowly by consoling through a clot and there by starting the fighting process to remove the external particles, organisms and finally bring about the same structure with which the original tissue was formed looks to be a simple process. However, there are billions of cells, neurons and signals through out the body to make this happen and that without one's knowledge. Hard work and Pain are proportionally related. The ability of one's body and mind to get aligned towards a goal, and produce results may not happen without hard work on a daily basis.

Beyond a threshold, Pain and Endurance may be the only two things which will provide a healing touch to cure a multitude of illnesses.

The physical fitness test for a cop, for instance is a great endurance one has to go through if appearing for an interview. It includes a large number of tests which may be life threatening, however the routine conduct of the tests results in one's endurance towards the needs of the job. Delivering work on a daily basis generally results in fatigue and boredom. This is because the mind over a period of time refuses to think further in the same field and needs to renovate itself. Artists and Musicians periodically face this problem when they try completing a task on a time bound basis. A goal-setter trying to train one's body and mind for a race or a tournament, generally wants to reach position one, with very minimal time and training. Unless, the goal setter knows many of the tricks to perform it as a magic, it's almost next to impossible to achieve this. How does one even measure to show if there is a level of hard work on the activity he or she is intending to perform?

We can take an example of a student appearing for a competitive exam which needs heavy memory usage such as remembering Mathematics formulae or Equations in Chemistry. The student may have to read through the various equations and formulae many a times in order to reproduce the visual images during the exams, in strenuous conditions measured with time. Let's presume different students as characters A,B and C portrayed through a small play shown below.

Character A   Character B   Character C

Though all the three characters portrayed above are born same, with same Mass and abilities, during their course of growth the characters get to meet and greet another set of different masses, with either equal, greater or lesser potential. The potential here is measured as a unit of perseverance. There is a subtle level of energy leakage between the fictitious characters mentioned above and the masses they meet over the years. Let's say the characters are at year 9 and able to grow normally under various conditions physically, the idea of education is to measure the level of perception by the character. Hence comes a point, to justify the need to have different exams and suggest the right course of direction and momentum needed for character A, B or C in their later years. One may think about the point we are trying to analyze by comparing the three characters with PAIN. The three characters portrayed here need not alone be compared in the field of education for succeeding or failing painstakingly, they can also be compared in their own fields where people are competing on a daily basis. Assuming that the shown set of characters are fed with the same food, allowed to work independently in the same environment and not having any physical and psychological ailment, the testing procedure will hold valid based on certain criteria. We can analyze the paths designated and movement towards the specific target through a simple race diagram and the position of each character at different points. For the sake of clarity, Character A is normal with solitary characteristic, Character B classified as a lead potential with a group of similar minds and Character C, a holy natured one and attracting minimal distractions.

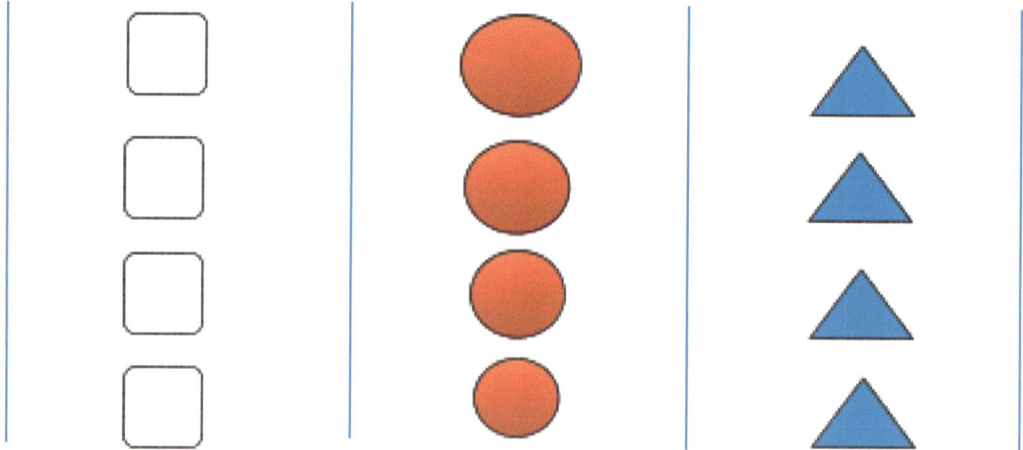

As we see in the race diagram, the character B clearly gathers higher energy levels due to larger potentials and this may be attributed towards the classification done already beyond year 9. The Triangular character C, with a holy nature exhibits the ability to be same after any number of years, whether excelling in the testing procedure or not. The character A now, in the race diagram slightly shifts towards the left since character B and C have an affinity of maintaining a group. Character A, does play an important role, since the character can potentially exhibit the energy needed at a later point to become character B or character C. The character race diagram provides a good inkling, that may point us towards an untapped potential in each of the characters, possibly the goal-setters who may have to incur pain to achieve one's vision. The only difference here may be that many of the inventors were exhibiting characteristics of B, early in their lives and were able to succeed by showcasing their work at a later point in the race compared to their counterparts. They were also able to do so persistently showing great levels of patience and perseverance with incessant levels of energy derived from their counterparts in a healthy competition. They finally would leave the results to the general public to accept their work, and in case the work is rejected then try again and succeed until they see tangible results.

A competitive race, in fact provides the needed momentum discussed in the initial part of the book to derive the energy and create an inertia there after. Greater, the initial momentum greater is the level of energy needed to offset the inertia created by the momentum. Inertia can simply be compared to the force produced by a body or group of bodies inside a flight after it makes a landing and applies brakes in the run way. In lighter vehicles, levels of inertia is lesser since the initial acceleration is less and speed is moderate. However, in a plane the initial acceleration to lift the heavy weight is huge and speed is high. This results in a large inertia on the bodies inside the plane which has to be offset using a large number of masses with opposite potential or a travel to be completed in different directions in which the inertial energy gets disbursed. This asserts the fact that the needed energy for reaching the goal exerted by most goal setters would be larger than the needed force, because of the fear factor of not reaching the goal. Hence one can derive the equation, that most of the goal-setters find it difficult to control the inertia after reaching the $N-1_{th}$ position or may be the N_{th} position itself.

One may see that if the paths of the characters are interchanged in the due course, it would result in a positional shift and a centripetal force due to exertion of inertia left over by the previous character's work activity. The exertion of force towards one's goal by anyone is not only dependent on his or her own work and efforts, but also depends on other's activity. For instance, if the force applied by character B after track change is higher, this would invite Character C to follow character B and momentarily the levels of energy dissipated in the same activity is higher and would pull character A towards the right border line. Hence, the law of natural force will automatically apply to reduce the energy levels and hence provide an inertia at a point which may allow or not allow the characters to reach $N-1^{th}$ position. One may also see that the levels of energy dissipated or needed to dissipate as one climbs the ladder towards the path, is highly time consuming and provision of necessary tools such as Augmented equipments to reach the position becomes vital.

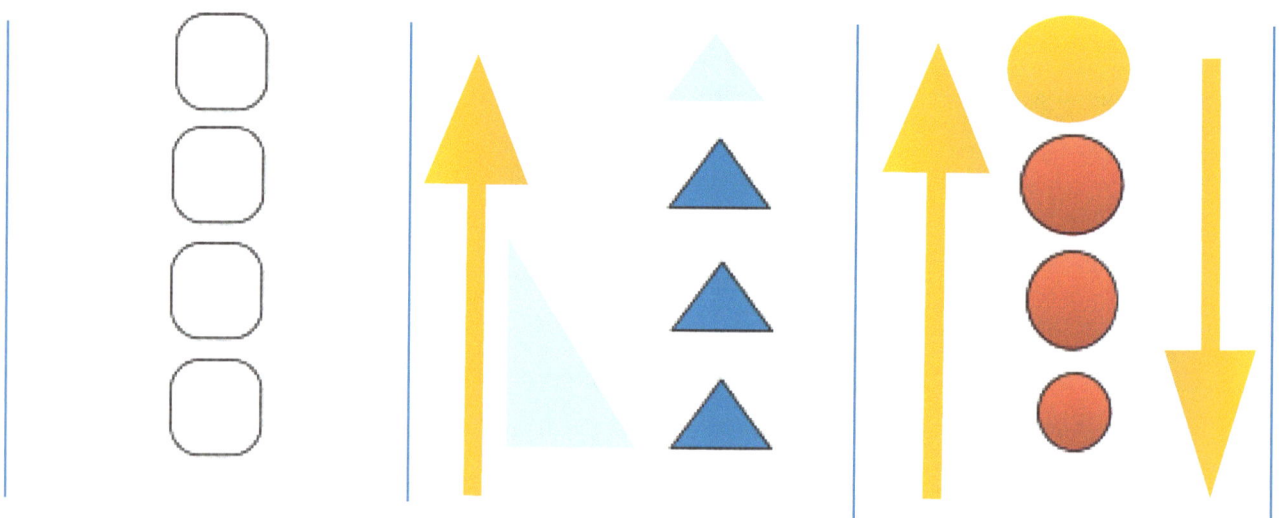

In a Modern Day environment, there may not be anything named 'GOAL' to achieve, since the law of natural forces will offset the energy exerted by multiple masses towards the same GOAL in a large population with increasing education.

Instead, the larger masses may join hands to walk towards the target slowly and provide results as instructed by a global balancing force.

The magnitude and direction of the global balancing force may or may not be decided by the smaller masses with higher potential around the globe, since each of the larger masses may slightly align towards the border when needed.

Energy exertion by these fictitious characters, has a level of pain associated with them. If a person working in a factory is asked about the pain point, he or she may quote the necessity to stand and work all the day long, when compared to an agriculturist who has the need to be athletic in ploughing the field and may whine about the necessity to move around to fetch water for the fields. We are talking here about two types of pain. One is a physical pain caused to the body due to the work exerted in a factory and Two, a pain exerted psychologically to the agriculturist who is dependent on certain external factors for completion of the work. An idealistic procedure to mitigate these two scenarios and help both these examples succeed in their vision of moving up the ladder would be to provide necessary education in the first place in their fields, clarify resolution incrementally and gather inputs to help with al-round support at equal intervals. Unfortunately, in a real time environment with large number of masses, the levels of attention provided by smaller masses may not move the examples to reach the successful position or even provide the right momentum and direction towards pain mitigation.

For a Goal setter in the field of education, one needs to be extremely cautious, since there is an upward surge of energy towards the meeting point of the complex neural networks in the body. This also means a large amount of reactions taking place in the mind than the reactions equally spreading across the body towards the other parts. Equal intervals of comprehending the subjects and reproducing the data during the testing scenarios is viable than inducing a great deal of upward flow of energy to produce tangible results. It's also interesting to know the fact that the flow of energy towards the top portion of the human body induces a movement of large number of healing bodies such as platelets, Red and White blood corpuscles because of large levels of energy spent for activities such as reading, comprehending and reproducing resulting in an oxidative stress.

A good example to quote would be the body's response to many of the activities performed on a day to day basis. If a specific part of the body is used extensively, it needs to be supplied with rich nutrients to reap extensive benefit or it may become vestigial.

'Vestigial' process in the field of human-evolution plays a crucial role in determining the retention of many of the characteristics shared with one's ancestral origin.

The element that is extremely important for human's to breath, may just be elicited with ease, by a fish swimming in water.

To be precise, its sagacious to identify one's strength not only in terms of physical fitness or psychological ability, but also through the vestigial process in which the human evolution has applied itself on the characters at play. The luring potential in the field of education and various other fields available in the world helps to produce a latent pool of talent and healthy race for a survival which would result in the fittest coming out victorious.

The below diagram shows a representation of interaction of different masses caused by the external forces such as centripetal and centrifugal acting on them. If Earth is assumed as a Mass and other planets assumed as different masses, the various effects of these masses on Earth and, Earth's reaction towards these masses can be understood using the below diagram. The initial path shows a planet such as Earth resting on a triangular body which keeps the Earth look stationary, though there is a constant force on Earth provided by various other planets to keep it spinning. We can comfortably say that the

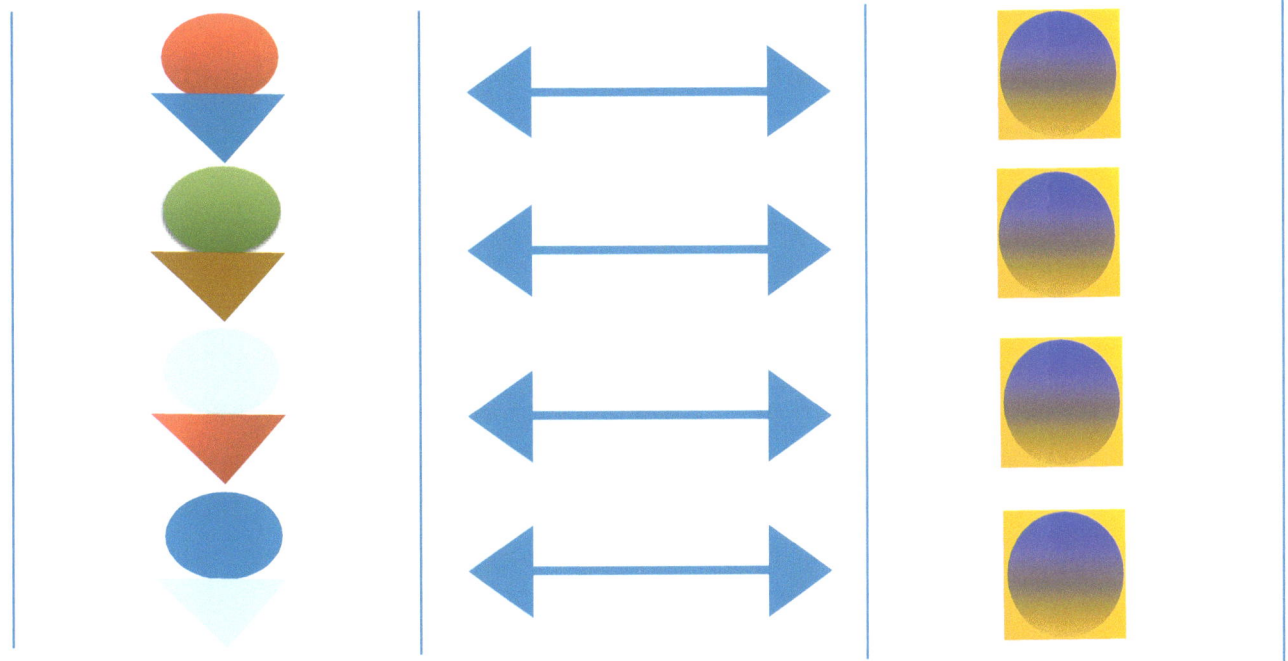

Earth as a mass is restless, though from a bird's eye view it looks stationary and calm. The forces that may be attributed includes the forces of the nearest assumed planets such as Venus and Mars. The other forces from different planets, stars and satellites are ignored in this case. When we try spinning a body of mass M (assuming it as a conductor) in a Magnetic Field, F' then M produces electricity. There is a level of attraction or repulsion based on the polar affinity of Mass M and this principle is used in many of the modern day applications and even in the field of levitation of a huge high speed trains in some of the countries. Once this is applied to the spinning masses such as Earth, Moon and other planets, then we can get to know that the Earth to keep the living beings stable is rotating and revolving at a speed which is of tolerable limits with the help of other planets, satellites and Stars. If one of the planets or stars closer to Earth is revolving at a larger speed or slower speed, it would have a profound influence on the Earth's magnetic field and Earth has to acquire characteristic of the other planet or the star which is trying to create an "Influence" on the Earth. The specific reason we are discussing all about Earth and not about other planets and stars is primarily due to the Goal-Setter's position to achieve his or her goal under the influence of Earth's gravitational field. If the Goal-Setter is made to undergo the same exercise of stepping towards the goal, in a satellite such as Moon or planet such as Mars, the calculations and time taken towards this activity will be entirely different since digestion, eating and resting takes longer in many of the other planets compared to Earth and will have an effect on time taken to reach the goal on the other surfaces.

Reverting to the diagram in question, the momentum exerted by different masses to each other while reaching the target, has a great effect on the globe when the mass is small but the goal is huge. We can assume that the Earth for instance is resting on an equilateral triangle, which is an imaginary Force F'. This imaginary Force is the angular momentum caused by other planets, satellites and a rotational inertia caused by Earth because of its revolutionary and rotational speeds. Unless, the rotational speed of Earth is considered a variable, which keeps changing to manage and keeps itself in the orbit, the triangular imaginary force may make the Earth slip from it's orbit and hence varying the ability of one to reach towards one's Goal. Think about a toy designed to rest on a ground with rotational inertia because of spinning. What keeps the Toy spinning is the external force created by the Toy called the angular momentum, which ideally is moving towards the center of the object because of the polarity created due to center of gravity and hence becoming a centripetal force. This makes the toy to spin constantly even though it has a bigger head and a smaller foot to balance. If Earth is considered in a similar fashion, there is a large angular momentum and centripetal force acting on this towards the core and creating an imaginary force named F' to balance itself. The triangular force considered, if made to touch four quadrants of the Earth's surface would result in a transformation shown on the right, in which the Earth absorbs much force needed for the sustenance of the spinning and gives away the remaining force as rotational inertia for the Earth to continue spinning. This also helps the other planets and satellites to remain intact in the due process. However, *at any point of time if there is a credit of force owed to the nearby planets or stars, then the Earth as a mass starts exhibiting the characteristics of the planet or the star to which it owes the energy to be delivered back.*

Though there are various forces which may work for or against a goal setter's path, the ability of the goal setter to move ahead in the designated track like a stallion will provide a good yield at a later point. This adheres to the principle of physics of action and reaction, instead of uncertainty. The only point to fathom is to see whether the goal-setter is able to complete the ride in the fly wheel, which when rotated over multiple times start producing results at a particular point without the knowledge of the goal-setter.

The left part of the may be compared to a fly wheel and the right part can be compared to the object of GOAL in question in the above illustration. The level of Angular momentum needed to produce the rotational effect on the GOAL to be achieved, can be gigantic or minuscule in nature based on how and by whom it is handled. There are certain goal-setters, who do not need an angular momentum to achieve the vision, instead work well with other goal setters to inch towards the position. The direction in which the largest energy gets dissipated or the magnitude levels at which the gyration decides to let go of the particle nature of the GOAL is not so easily predictable. Hence the Uncertainty principle is always valid, that said it would not be valid to hold on to the theory of uncertainty always to push one's goal towards the side since the confidence level of the Goal-Setter and the Goal-Setter's environmental companions may dip to a lower level through a negative wave. The field in which one has to shine has to be decided well in advance and if in case the number of fields available to play or perform is dwindling, it would make sense to allow more inventions to line in place so the field becomes wider for larger audience to participate.

On a final note in this topic, **Predictions done by great inventors did not happen at an accuracy level of 100%, since the predictions are in the air and well understood by other players who have the potential to disrupt the prediction.** One should verify the facts well using applicable knowledge, research them with enough time and then move on with judicious step towards the goal with sincere devotion in the field instead of being blind folded towards accepting any prediction on par.

VENA CAVA

Its fascinating to know that only two of the veins amongst other named veins in the human evolution, have the capability to carry pure plasma to the organ named heart. One Pulmonary Vein and second Umbilical Vein. Carrying pure plasma to the necessary organ is the key to survival since Heart as prescribed, remains a key organ to keep the pressure up for the other set of organs to function well. There are a huge number of veins and arteries in the body, which has an important function of transportation of a good plasma, possibly the universal healer. Now, the transportation of impure matter and pure matter, both play a key role in the survival for a good living since the impure matter has to be purified through some mechanism. The mechanism of purification can be easy or difficult and may take a long time in few living organisms. This also depends on lot of factors such as external environment, purifying organ and ability to transport the impure plasma. Let's presume if the organ which is to create pressure for transportation or the transportation canal refuses to coordinate, there will be a stagnation of impurities and the specified organ becomes a feast for anaerobic micro organisms. The term Vena Cava refers to the group of channels which are used to transport the set of designated impurities to the purifying organ and also the purified smear to the same set of organs for renewal and replenishment.

In an environment where there is minimal pressure to pump the nutrients, or minimal levels of nutrients available with high pressure, there beckons a process of treatment and healing. A goal-setter needs to have certain set of conditions before thinking of stepping into the tracks for competing. The venous system, the aim of which is to transport impure smear to the organ for replenishment, has to co operate well in advance before the goal setter begins to start his work and activity. The higher the energy spent by a single organ in a living organism, more the energy elicited from other organisms in which case the ability to heal depends largely on the environment instead of the individual. This has a profound effect on the transportation channels, since the percentage of impurities transported increases and the efficiency to transport nutrients for replenishment decreases resulting in an overload. If the Goal-Setter works continuously towards the path to achieve the goal, whether monetary or non monetary, the set of organs needed for completion of the activity are put to use most of the time. This means that the time needed for renovation, decreases and the body's innate mechanism provides a signal for the goal-setter or the equipment he uses to reduce load. This can be compared to an overloaded machine in a factory handled by a lot of staff. Unless the machine is handled by the technician who has spent a long time with the machine and knows the intricacies, ways to handle the parts, technology used underneath and cooling period to be provided, it becomes next to impossible to reduce or eliminate the wear and tear of the machine.

The reason to touch upon Vena Cava as a topic in the midway of this book, it to make one realize how well the body has undergone a great deal of an evolution to nurture, replenish and undergo transmutation. It's also interesting to know that at a point of time, the umbilical vein which is used to feed a baby from a mother's womb in a neonate, is not taught to a carrying mother but already taken care of. This is applicable across living neonates and need not be researched too much. **Most impurities, which a mother thinks of to protect the developing fetus from, is already assumed to be subsumed in the knowledge base of the nervous system**. If a Goal, has to be induced in the nervous system or the sub-conscious of a child at a very young age, it means that the provision for inching towards the goal had commenced when the fetal development started or may be even before. Once the growing goal-setter, shows interest towards a particular field of arts or education the best thing that might be needed is to give a helping hand and set a strong foundation to avoid distractions.

There are lot of instances in which a Mother and a Child have shown great interest towards Education and were able to successfully achieve greater heights together.

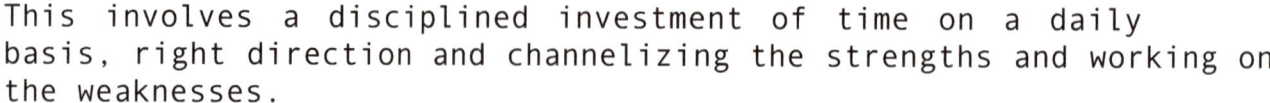

This involves a disciplined investment of time on a daily basis, right direction and channelizing the strengths and working on the weaknesses.

The above two points taken care of, mostly results in fixing of the venous system in a gradual way since the laws of nature prescribed before, will work on healing the body and mind, just in case of exhaustion and desperation.

The context of Vena Cava and Pressure created by a neonatal body to balance the organs to function normally is a topic of research. If the Goal-Setter for instance has a high pressured job to bring an ambulance to save a life, the pressure for the goal-setter is quite high and body condition is mostly in duress. Since part of the Vena Cava is used for lower and part of it is used for upper portion of the body, a well advanced neonate already knows that the angular alignment towards the right or left side of the body results in betterment and deterioration in an alternative fashion. Some neonates generally decide to ignore the conscious alignments and allow the subconscious nervous system to take over to adjust itself in solitary conditions which still holds to be a valid step towards healing and renovate to start working on the activities again.

During my young age, I found it physically challenging to be in a race. Instead, I did find another interesting way to keep the mind occupied. What a competition does to you as a human is to improve your senses in a way that other lower beings may not think of. It allows the 5 senses, to balance and provide inputs to the thinking hat in an ameliorated manner. It also helps your body to sharpen the senses which are more useful, through vestigial process of evolution. If for instance, what we see to perceive and comprehend has a lot of augmented reality images which goes beyond the 3 dimensions, then an upsurge of plasma automatically takes place to continue the process of perception, with more focus on eyes, nose and ears. The part of the Vena Cava, which is supposed to clear the impurities gathered during the process has to perform an extra duty during these times and these Competitions in a way, train the neural networks to face the challenging world. The term 'Rat Race' in English has it's roots fixed to the same instinct one gets towards reaching the destination. Rat Race is a common term when Rats are allowed to compete in a race, with nothing other than cheese on the other side. Rats, having the perception only through the nose, get into a hasty competition for the food source and hence attracts the predator's attention and so on and so forth.

A blind rush in a competition towards any destination or a single goal in a large crowd or population, would only result in a rat race. If the path is well set before beginning to move towards the destination, will help ease the process of a competition and also make the goal-setter's health better. The way in which this can be achieved is through training the body and mind on a day to day basis. For a beginner, who has larger physical needs with higher psychological ability, the matrix would skew towards a moderate time to be spent on a daily basis as compared to lesser physical need lesser psychological ability. This also depends on the field chosen by the goal setter. **It may not be a valid game for a psychologically and physiologically stronger person to compete with a physically challenged weaker one**. The matrix diagram may provide us a pointer towards the overlap of a field chosen such as Arts, Sports, Education directly applied to the BMI-Intelligence Quotient of one's potential. The BMI refers to the Body Mass Index and Intelligence Quotient refers to the intellectual ability of the Goal-Setter. This also varies with Time and Speed, since both directly has an effect on the BMI-Intelligence Quotient in the chosen fields. The Matrix can only be used as a tool towards the betterment of the goal-setter, which of the areas have to be concentrated, how to produce better results at ease and improving the Temperament overall. It need not be used to produce momentum towards reaching the goal since most of the idealistic goals have a residual value associated with them which can be leveraged. In other words, the matrix tells us how much is the rest to be provided to the body for the mind to start thinking.

The matrix and mosaic diagram shown gives us a glimpse of one's Goal, Path and the effect on the nervous system. The black shows the impurities to be cleaned up to improve Intelligence Quotient, Brown shows the Improvements to be done for a physical fitness and then comes the way for a Goal-Setter to start inching towards the path with Year on Year improvements in a controlled environment. As we can see that in the Year 1, there is an yellow region which signifies the Goal-Setter's interest in moving towards the goal in the first place. This creates an aura around the environment of movement. The movement without much substance, just creates a void and brings the goal setter to Year 2 in which the purification takes place either guided by a master in the same field or by an eternal force if there is no master, but the stead fast movement of the Goal-Setter with Sheer knowledge and eagerness to learn further. In case the Goal-Setter fails to take the guidance of a master, it only lands him or her in the year 3 with no-space for BMI-IQ measurements, or an exhibition of the latent talent of all the perceptions gathered. If the goal-setter is shrewd to gather a large amount of data and put forward this with good acumen, it results in an earlier opening of matrix in Year 4, with the lower portion as green signifying the ability to perform well physically and the top portion being blue, signifying the ability to showcase the inputs gathered at ease.

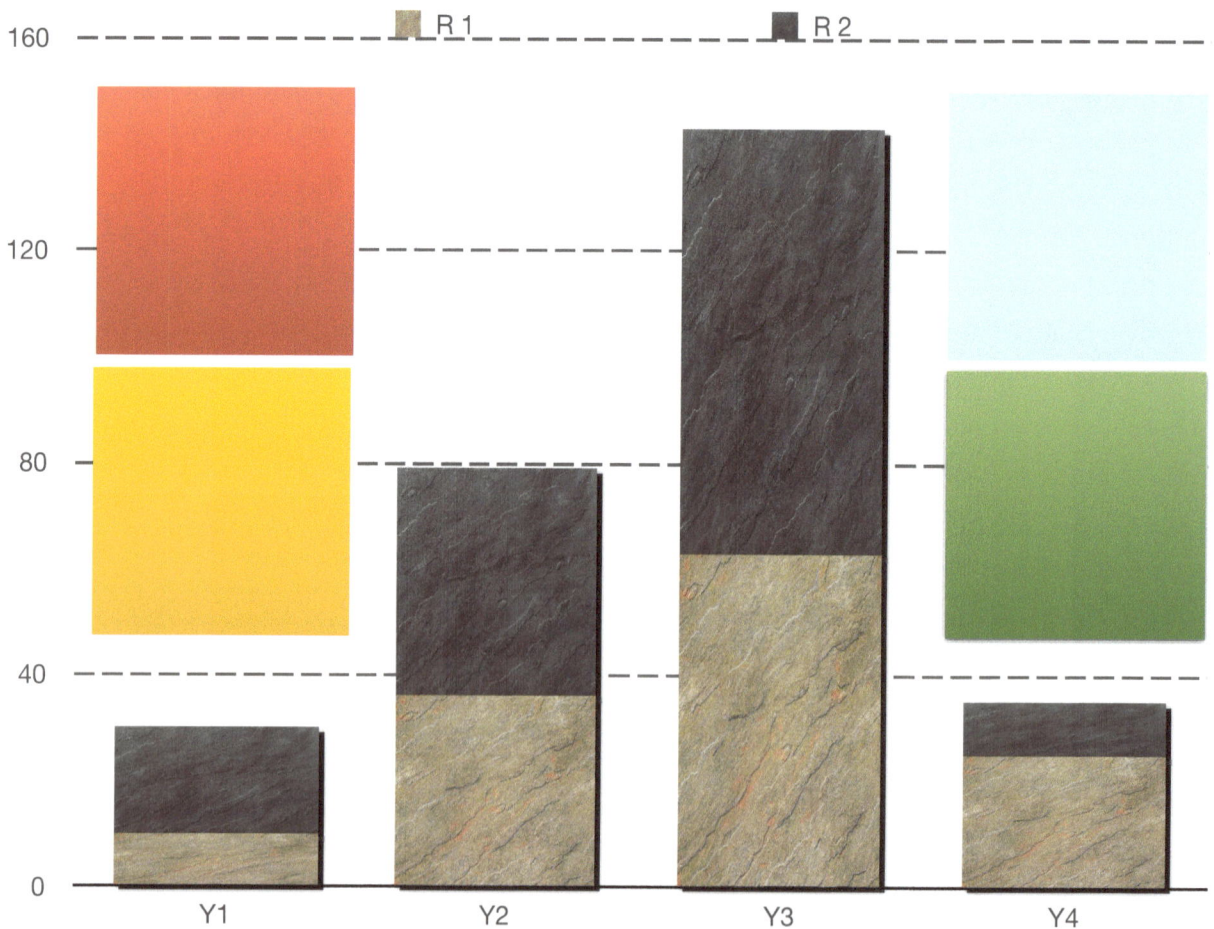

The introduction of a matrix to guide one towards the Goal and The Path to be chosen may look to be another Augmented Reality Program with manipulated co-ordinates. However, a theoretical subject with dry areas covered extensively to educate a goal-setter would just look to be a house fly trying to reach the stars. The various educational aids introduced by the inventors have created a revolution in the field of Arts and Education. It has also helped them nurture and protect the invaluable souls, providing them the needed seclusion to achieve greater heights. One may wonder on what can be an inspiration for inching towards the activity needed for succeeding in the goal, or even completing the first step towards the target. For this, one has to listen to his or her own body and mind. This is an important point for observation for the reader, since it tells on how to provide rest to one's own senses and still use them to think well using the nervous system. I am quoting these three points, to avoid confusions on what makes an art completely different from the other and how to overcome obstacles during an art creation process which can be showcased as a complete novelty instead of an output of a copier.

Most, if not all of the artistic work in this world are derived from what a goal-setter had perceived earlier. This may be a piece of work in Music, Arts, Medicine or Physics.

Novelty, in a way is about the introduction of a unique proposition which will attract the audience to understand complex theories and equations in a simplified environment to provide a healthy comprehension.

A Copier machine for me, is a novelty though it does only nothing other than Copying content. A world without a copier machine may have resulted in inextricable complexities to establish an identity proof for doing anything and everything.

Though Copier Machine may just be a novelty, using copier machine to produce work without efforts would result in anomalies. This can be portrayed using a binary tree in an agricultural field. If a tree which has the ability to multiply itself into two without any differentiation and value add, then it would occupy the entire field in very less time, the field that can be used to feed livestock.

If the binary tree shown, is considered as crops, then the copier is considered used for good work and instead if the binary tree is considered as weeds, then the copier has to be kept in a controlled environment to limit the population. The copier in this context, neither has an artificial intelligence to differentiate between the two nor it can be provided an added intelligence to do so, since the copier may over rule the capability of multiplication of anything and everything it happens to encounter, since there will be very minor difference noticed between these crops, weeds and other crops.

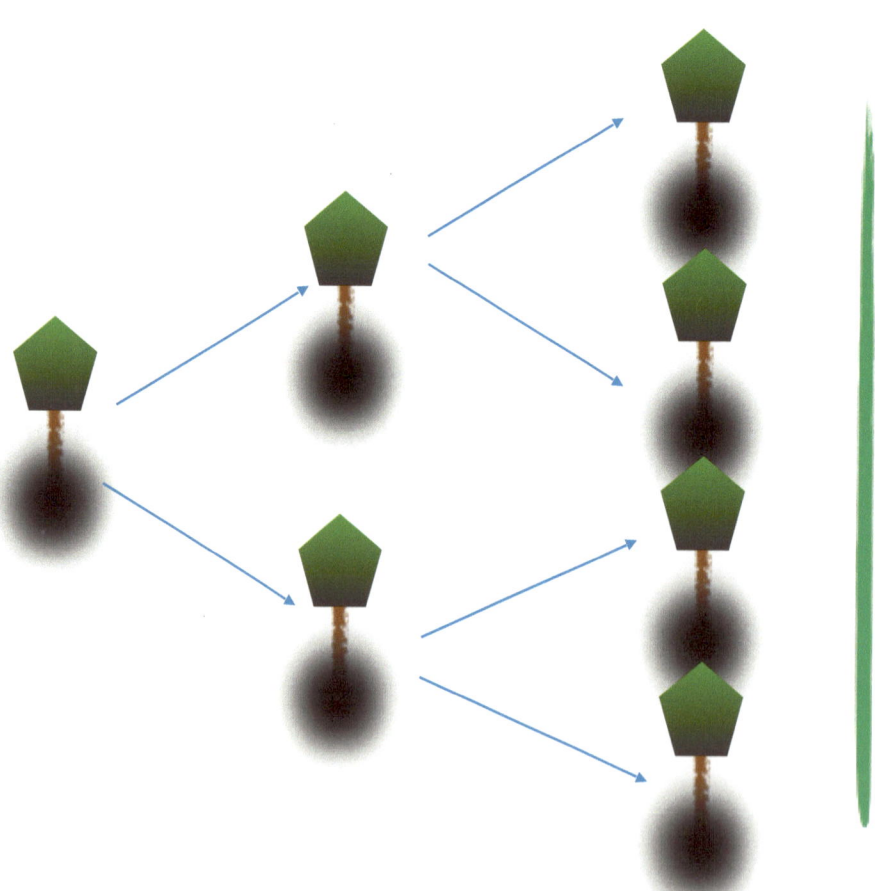

The binary tree has its roots, even before many of the inventions came into existence. Right from Computer applications in storing data of customers to establishing an identity of an individual using a finger print match etc.. One may use many of the applications of a binary tree in moving towards one's GOAL. We can analyze this by providing a little more explanation about the binary tree concept and how it can be used beneficially in one's path. A binary tree can be compared to a cell mitosis happening in a neonate on a day to day basis. This is applicable from a simple plant which grows to good heights over a period of time by consuming water, to a neonate mammal which grows over a period of time by ingesting food on a regular intervals. "Setting of a GOAL" in one's life beyond a particular age, provides a platform to the mind to work on tasks at hand instead of energy getting spent in a direction which may not be fruitful. Once the mind gets a platform to work on, the speed of body's cell division reduces to an extent whether the division is positive or negative. In an ideal environment, one may not be able to control the direction of the cell division, but the speed of cell division can certainly be. The ability of one to stick to the decided goal, provide quality value add at every point and get it to the level of delivering the artifacts as a complete package sets the body clock in the right direction, provides natural climate for flourishing and replenishment and in case of a multiplication in different direction, asks for proper rest.

This takes us slightly backwards towards the venous system and its effect on the Goal. The cell multiplication speed directly depends on the speed of the plasma flow towards the organs at work. The rate of cell division results in growth and the quality of cell division depends on the nutrients. If a plant is fed with necessary nutrients that the plant needs, the results can be seen with naked eye. Apart from the necessary environmental conditions, the plant may show us on what Goal it has towards the growth. For instance, a Cactus plant may survive with very less water, however a coconut tree may not. One may ask if a plant has a similar venous system like humans, answer may be No with a slight twist that the plants are mostly dependent on the humans for certain processes such as photosynthesis and transpiration. The Goal and the path set by a Goal Setter, may not work well in an atmosphere in which there are not enough nutrients for body and mind or the pressure given by external forces are very less in reaching the target. Just in case, if too much

of a pressure is given with less nutrients it may result in a detrimental health vis a vis,
very less pressure with lot of nutrients. A right combination of Speed of comprehension and
Quantum of knowledge fed on to the both mind and body at calculated intervals may work like
magic, in providing the right momentum in taking one towards the position aligned with a
moderate percentage of certainty.

The Inferior and Superior Vena Cava in the nervous
system of a neonate shows the natural alignment of the body to eliminate toxins while working
or resting. The term may refer to Inferior or Superior, however the importance of both seem to
be equal. Inferior tries to work on the lower portion of a neonate while superior tries to work on
the upper part of the neonate. This means both have to complete the job simultaneously on a
single pressure pump of the natural pace maker named 'Heart'. The Pulmonary and Umbilical
Vena Cava, with a duty of transporting the good plasma to the necessary organ may be an
exemption. One may still wonder, on why these two veins need to be an exemption when there
are lot of arteries designated for the same purpose. Answer may just lie in the size and vestigial
evolution of the neonate. Over a period of time, the necessity for the survival of the neonate
would have initiated couple or few of veins to undergo a transmutation and may have joined
hands with the arteries to improve the process of purification and give the innate feeling of
protecting the child from impurities.

One may also think of this as a pure scientific phenomena in which there is a co-operation amongst the neonates to protect the food. At a point of time, when a mammal gets food after a very long time, then the nervous system of the mammal decides to protect the food instead of consuming it. It may be treated as a rarity of an event in many cases, however it would be a naturally occurring phenomena in mammals and neonates. We can also see this with an example of how the body reacts for different types of foods provided at different time intervals. As shown in the image, when the food is provided to the body at regular intervals in a tolerable limit the body accepts the food and digests, however when the food is provided to the body at distant intervals though with a variation, the body tries to protect the food instead of consuming which is a process used by hibernating bears in winter conditions. This also goes on to show that for survival of a distinct race, one of the neonate in the race will decide to protect the food for the other distinct neonates in the same race to survive and flourish. This may be through the mother of the neonate which protected it using the umbilical vein or father or sibling of the neonate which gathers food and makes provision for the survival and growth.

Goal, by and large is a consolidation of many smaller goals fed to the mind on regularity. The consolidation will happen only after a large quantum of energy is dissipated towards the goal with division of labor completed exquisitely. The larger spell bound picture, which the audience may see after the performance of a goal-setter is complete, will mostly be through a strenuous, rigorous and meticulous work activity practiced and performed in a disciplined environment guided by an external force and a healthy environment. **The first step one takes towards the target decides the trajectory, which can be compared to a rocket soaring to great heights within minutes and following the trajectory till the journey is complete.**

6 STIGMA

Stigma and the process of Evolution are distantly connected. One may wonder, why certain attributes are inherited to her or him from a father or mother or a peer or an environment. The process is termed as Dominance and Recessiveness of alleles during the make over of a neonate fetus. This generally results in a stereotype (behavior) and uniqueness of one individual (human) or a plant compared to other plants. Let's say if we have only mangoes as fruits available for consumption or carrot as the only vegetable available to be mixed with the staple diet. The result may be atrocious. This also takes us to an interesting topic of inheritance in plants and humans. An extensive research has already taken place over the last few centuries, on certain types of stereotypic behavior with certain kinds of plants and humans. The dominant gene is something which expresses itself with higher confidence level compared to a recessive allele. If there was no dominance and recessiveness amongst the subject of study, then what would result is the multiplication of cells to an abnormal level as shown in the binary tree in the previous sessions. Fortunately, what had happened in the theory of natural selection was to provide a platform for individuals to portray the dominant alleles at various intervals, so the alleles with recessiveness can absorb less, few or most of the dominant characteristics.

The capability structure of individuals with extra-ordinary traits gets either reduced or capability structure of individuals with recessive traits gets increased with the natural selection process. The term Stigma generally refers to the distant behavior of an individual or a plant due to specific attributes which are considered as disgrace or different. This is due to the fact that, every birth as decided by the eternal force is completely unique. Though there are theories that individuals can look very similar, take for example the process of twin brothers or twin sisters who are born alike. There are minor characteristics such as finger print or the chromosome structure which makes them different though other things are common. Today's science has improved to such an extent that, the analysis of dominant and recessive structures in the genes, has become a routine in providing direction towards a healthy living, apart from playing a role in forensics and identity provisions. Most illnesses as per the stereotype theory is mind borne and can be healed with improvement of confidence, which comes along with the fixing of physical ailments. There is a research subject on deriving the age of a tree using the annual rings in the wood. This is an interesting point to check on the wellness of the tree, the products of which is used for consumption as food and medicine.

Stigma can have a great effect on one's behavior and work output. In a typical environment of growth, the stigma looks to be having reduced effect on one's well being provided there is enough room for a healthy vision and goal setting. This can be verified with an example that we can experiment. Growing a plant with adequate sunlight, watering on a daily basis, manures supplied for assimilation of nutrients and similar plants around for mutual sharing of nutrients. If a plant is grown with minimal supply in any of the above choices, then it directly affects the life of the plant through various means such as pests, drying of leaves and stems or even destruction due to external forces due to vestigial process. Social Stigma is a specific subject of interest, one may read through if needed to understand the nuances of how well a child can adapt to the Goal with building social pressure with various short comings reflecting on the body and mind as a result of interaction of dominant and recessive alleles or the environment in which the child may be competing. One may not be able to avoid social stigma and work in destitution completely since it would be straight compared to a competition without competitors. In most cases, the tapping of unnoticed potential in a child only exhibits when allowed to perform, play or compete in various fields. Once the diversionary technique is applied on the child's mind, the stigma may not have a detrimental effect and may start resulting in useful work for inching towards the Goal. The Path mostly covered with difficulties would start looking to be the stepping stones for success and once the child starts showing interest, the Goal becomes an area of interest for other set of competitors in that field.

In a typical factory environment, the item to be manufactured is produced through a conditioned labor, contributing time and efforts over a period of time. If we presume that the factory is a tooth paste manufacturing company, the parts to be manufactured for the tooth paste such as tooth paste cap, tooth paste body need to be produced through separate machines as compared to the making of the tooth paste itself which is made through human intervention. In a large scale production environment, the manufacturing process becomes laborious and hence came the inventions of machines to complete the process in a simplified way without consuming much time. The trade secret in making of many of the fast moving consumable goods such as tooth paste, hair oil, cosmetics and soaps remain to be a critical factor with a value add from the ergonomic design of the product in totality. If the "Mother of Invention" named 'Necessity' becomes reduced in demand then the trade secret takes a back step while design and affordability of the product takes the front seat.

The trade secret is a process using which most of the modern companies flourish and help transform the mankind. Example can be a simple cool drink which has a value add such as a vitamin, and help replenish the thirst of sportsmen and women during summer season. The trade secret is generally mentioned as a clandestine, in order to maintain the brand of the company which manufactured and introduced the product to the market. **This is also adopted to avoid a copier effect and a trade conflict between nations.** For instance, the natural resources of a particular country may support the production of the specific product where as the natural resources of the other country may have already exhausted the reserves. This remains a critical aspect in fixing the price of a product with the trade secret uncovered only for the board which had set up the manufacturing unit, instead of a population at large. This also necessitates the nations to accept and reward the painstaking efforts to work out an optimal plan to co-operate and export or import the products as and when needed in their nations, for the betterment of mankind. One may still question the relation ship of a trade secret with the life of a Goal-Setter. Few extra-ordinary works such as Arts and Music are preserved for a long time to energize and renovate the thinking of humans and make them manufacturing giants and hence the Goal set by a child at a young age needs to be monitored at regular intervals.

Name	Ergonomic Design	Trade Secret	Implementation Time	Endurance	Weighted Average
Product A	45	23	68	91	56.75
Product B	58	57	56	89	65
Product C	15	5	90	90	50
Product D	32	66	35	89	55.5

The above table compares the list of products, A B C and D, to be manufactured in a factory environment. All other raw materials are said to be on par, we are trying to compare the output produced on the basis of 4 parameters. One the ergonomic design, two the trade secret, three the implementation time and four the Endurance.

Ergonomic design refers to the comfort level with which an end user can start consuming the product. For instance, a tooth paste which is difficult to consume on a daily basis though the product is extremely good or a tooth paste which is extremely volatile and becomes a liquid too soon because of the design may not go well with the users. Trade secret, on the other hand gives the product a monopoly to start selling, even if the other set of parameters take a back seat. Pharmaceutical companies extensively rely on the trade secret for manufacturing and release of a prescription drug to the general market. Since there are many pharmaceutical companies with different labels, the trade secret has become even more important to rely on than the other parameters. Implementation time, depends on case to case basis on whether the factory is ready to import the needed machinery to produce the end product, let go of the opportunity cost needed to import versus the profits that can be reaped at a later point. Endurance is the capability of a product to show resilience during extreme climatic conditions. Endurance becomes a key factor for aeronautical engineers, since the product manufactured for many of the jet planes have to undergo an endurance test for a very long time, without creating a wear and tear for the product. The weighted average of these together, decides the position of the product compared to the peers and makes a product different from the other, be it an output of a car manufacturing company or a pharmaceutical industry. Clearly, the weighted average in the above table, for instance shows that the product B tops the segment, assuming that the product B is competing with the peers in the same industry.

These set of parameters fed into a 3 dimensional plot would reveal the position of the product amongst the peers in a global environment. One may try to feed the data and use it for analytics to derive a trend for the product across geographies. Some times, this can be used to produce a predictive curve for possible catastrophes and incidents. Let's say if the usage of few of the industrial products increases in proportion, the rising cost of hospitalization increases proportionally and if the number of hospitals with effective equipments are not enough in a particular geography, then the population in and around gets affected with epidemics. An interesting point to notice is about the Weighted Average shown in the graph. The Weighted average considered in this example, is a simple average of the four parameters considered during production. However, in large scale production across industries, and relating it to a national industrial output, the weighted average plays a crucial role in deciding the levels of input to be provided for each of the parameter, the levels of innovation needed in the trade secret, the Ergonomic design to be improved or the

endurance test to be prolonged.

Social Stigma and the process of 6 Sigma has a natural inclination towards generating momentum. How a book of a right choice in one's life, generates reactions in reader's mind, a product of great quality and endurance will share similar attributes. This is evident from the graphs shown and the demand created not only by the general public but also by the other set of players in the same field for understanding the endurance of the product without affecting the trade secrets, and encouraging a healthy environment. A Chief Executive or a Leader of a factory may have to understand that the Social stigma needs to be fixed amongst the minimal set of humans, running the show on a daily basis to provide the momentum to produce a work of great calibre. Social Stigma can only be fixed by renovating the path in which a Goal Setter starts walking. I still remember the physics laboratory experiments in which we used to generate sinusoidal waves using waveform generator and measure and analyze them using oscilloscopes. Let us presume that the path taken is a straight line instead of a sine wave, then it results in a linear momentum instead of angular momentum. A wave form path generally results in an angular momentum, in opposite direction to balance the mass.

Most mountains have an angular path on the road and rail constructions, instead of a linear path. What an angular path does to a rider, is to slow her down during driving and provide the necessary time to reach the peak. Some of the cable cars which allows the goal-setter to reach the peak sooner also have the risk of high friction and disconnection due to increased potential energy gathered within less time. In case of an angular momentum, this is reduced to a good extent. One may just consider this as an example while starting the journey towards the goal one has already set without oscillation. In a Linear momentum, the chances of collision increase in case of encountering other moving objects in the opposite direction, or even objects that do not move. In a movement generated by Angular Momentum Path, the goal-setter gets the ability to gel well in the field and manage obstacles and also gets the ability to move towards the direction needed at any point of time. **Social Stigma amongst peers may be justified with one specific point that the nature's way of informing the neonates to take an angular path towards the goal to help one another and provide support as needed for betterment of the system and reaching one's goal at ease.**

The term 6 sigma in crude terms, informs us that in a manufacturing process, it is expected that **99.99966**% of parts manufactured and assembled are to be defect free. This can be measured using DMAIC process, which is Define, Measure, Analyze, Improve and Control. One can very well understand the importance of Endurance while setting up an establishment or a manufacturing concern. The term six sigma, plays a crucial role in getting to define the industry norms for manufacturing a product for sales. Gone are the days when a monopolized environment existed for the products in which the availability of the end product and variety was very less in proportion. For instance, during the early 1900's we did not see a variety in the fast moving consumable products such as a tooth brush or a detergent. However, during the late 1950's an industrial revolution did create a level of willingness towards the improvement of products, quality and endurance. The industrial revolution has it's own effect on the products and establishment of factories. It takes us to the initial discussion of the conducive factors to be analyzed with the set of parameters shown needed to produce an initial outlay.

The 6 sigma and it's usage has definitely made a mark in the quality of life of general population in large. If we presume that the high standards of production is not set in a factory environment, in which a large amount of monetary investment has already gone in, it would result in a long implementation time and it would take a beating in the scores of the product percentage to get to a break even position. The levels of man hours needed to produce a trade secret seems to be an inverse proportion to the design of the product. For instance, some of the ergonomic designs of a pressure cooker which helps a household to produce food at ease and providing an alarm in case of an overload, is a unique invention when compared to a normal utensil used to prepare food. The time and efforts needed to think about the problem statement and produce a viable solution which may cure many of the illnesses for general public, would establish itself to be a trade secret and when this program is fed to a computer system to produce the item in large quantities, would establish itself as a design. When these two parameters are well taken care of, the implementation time needed to produce the item chosen, takes less time compared to an item without a proper design and a trade secret. The weighted average will play a role of differentiating factor or a unique selling proposition when the markets are unable to handle a monopolized environment and ready to take the heat of a multiplayer environment.

Social Stigma and 6 Sigma directly have an influence on the level of confidence infused into the reader's mind for creation of a product or a work of art.

In a population at large while considering a mass transformation, the Social Stigma takes a front seat to become the driver for the 6 Sigma process.

The momentum produced by a flywheel in making a Goal Setter move faster towards the Goal can be controlled using a combining factor of Social Stigma + Six Sigma.

Though there are many computer aided simulations now available to introduce a product in the market through a factory production, human intervention has always and will always play a role in initiating, continuing or completing the process. The measurement however is about the number of humans versus the proportion of machine work to produce an artifact. In a handicraft work, there is no need for a machine interference since the demand may be less, urgency slightly low and the whole intention of producing the handicraft work by spending time to produce a work of art may not be fulfilled.

The process of DMAIC introduced for the production of a product in large is applicable from spare parts assembly of a mobile phone to the spare parts assembly of a car. Let's assume that this is applied on a giant product such as an airplane, it means that the space needed for manufacturing this product will be enormous there by reducing the space needed for a normal handicraft work and to become diminished. Industrial revolution in a way has made a great progress in making the quality of life better for most humans, however has also resulted in reduced space, congested zones and depletion of natural resources. If we step back a little, and start applying the same Six sigma and a DMAIC process in the living room where we allow the mind and body to replenish, it would result in a different output altogether. One may think of this as less interference of machines and more indulgence of humans in which the genes of a neonate looks to be a better option compared to the output of a machine which produces radiation to a great extent.

Though I have showcased Social Stigma, a process which is more of biological in nature resulting in variation among humans and 6 Sigma, a process which is more of testing the nature of endurance and quality in a production factory, both these processes interact on a day to day basis to produce artifacts which may reach great heights or undergo a vestigial elimination. Few examples to ponder around may be the existence of heavy weight televisions for watching entertainment in big screens versus the light weight plasma LED television sets produced in these days.

Though there may be a crossing over of Social Stigma and 6 Sigma in the production of both these artifacts, other set of parameters such as the trade secret, design and endurance may take the front seat in driving the entire sales of the product, since it takes only close to a minute for any large scale manufacturing firm which employs a large number of humans instead of machines to take over a similar idea with a minimal variation and produce alternate product in similar lines.

The advent of social communities to counsel different communities of people, on day to day problems they face may only partly be a viable solution in a profit making industrial zone. For instance, the social community may provide a handshake on a day to day basis for betterment of the living being and may not be able to sustain the growth of weeds and byproducts created due to the binary tree reactions described in the passages. This can be verified with the growth of different nations till date, export import policies and depletion of natural resources of specific countries.

Over a period of time, the reactions allow the product to exhibit great resistance which decides the path forward. This can be explained with a classic example of how land mass was formed in the Earth with a ocean water hitting a hot molten lava for many a times and finally deciding to form an island which can be used for making a good living. For the land mass to get formed, innumerable reactions take place amongst the different natural elements such as fire, water and mud etc...resulting in a rock solid elemental mass.

On a closing note, this specific session aligns to the naming of this book, Mass and Choice - the MC and also allows one to take a relentless attitude of understanding one's Goal and working on the path assigned. Since the elemental physics always tries to bring out the uniqueness of the element in all ways and means, it would only be judicious for a goal setter to decide by oneself, if the goal set is unique and path taken is stead forward. May be at a point of time, when there is a huge level of fatigue and boredom, it would be wise to think again not to change the idea of letting go of the goal but to provide enough momentum by deriving the energy needed from similar fields of inspiration.

0 HM .7

Find the right answer for the question (qqqqqqqqqqqqqqqqq) from the below options:

a) a to a
b) b to b
c) c to c
d) d to d
e) e to e
f) f to f
g) g to g
h) h to h

The above example is just a glimpse of what a question paper may look like in the future with appropriate content filled for (q) and with apt choices. With improving aptitude of children and the increasing pressure of a school to perform, the standard of question paper may just touch a level in which it takes a very long time for a kid to solve a single problem, instead of a larger chunk of data getting into the child's brain and re-creating a reprographics scenery in the exam hall to complete the question paper on time. The aptitude of the children learning content and the aptitude of content creator, has to be synchronized at every point of time so that the school catches up with it's aim of having established a school. Let's say the school is for a set of students who are blessed with different sets of ability such as drawing arts or singing or playing keyboard, a part of the brain which is creative is working an extra duty to make them think in that specific direction. This also makes

one think about the possible reactions in the nervous system if one is not absorbing a term known as "Resistance" in English Language.

Many of the equipments which we use daily, such as mobile phone, computers and Televisions have a miniature mother board which allows electricity to pass through the components with a specific impedance to see what we visualize and perceive. This in the subject of Physics is termed as 'Resistance'. Resistance of a particular device plays an important role in allowing only specific quantum of electric charge to pass through the component needed. If the component needs a quantum of electricity 'V' then the resistance to be provided in the circuit board must be a calculus of 'V'. The resistance is measured by a unit named Ohm. The term resistance plays a great role in one's life in crude terms. If we allow everything and anything which we perceive through the five senses, then it adds a lot of input to the nervous system. This also at a point of time becomes indigested particles to provide access to free radicals. The free radicals in a body has a great capability to make other particles to become free radicals too, since what free radicals do to a body is create a craving to the activity, whether it is an activity such as eating too much, talking loudly or even a trivial but an important one which is always unattended to such as not breathing right.

An '18 day challenger' to let go of an activity **which is hardwired in the nervous system such as sipping a refreshing cup of coffee in the morning, needs an enormous resistance**. Without getting into an argument whether a cup of coffee is good or bad for health, the question in layman's terms is to see whether the Goal of letting go of a thing of high interest, can be achieved or not, with possibly the goal setter compensating the lost opportunity of working an extra hour with 2 cups of coffee every day after the 18 day challenge, on a lighter side. Let's see if we can derive a relation between the initial description of growing aptitude of children with a goal setter to show resistance towards an educational goal. It would make us touch upon one of the methodology of study named lateral thinking versus traditional movement across the ladder. A lateral thinker allows his or her creative aspect of thinking on a subject presented without boundaries. For instance, an equation proved to be a scientific phenomenon by an inventor, will generally be disrupted and questioned by the lateral thinker even if he or she knows that the judgement may finally end without much fruitfulness.

Unless, a question was raised by an inventor about an Apple falling towards the ground, we may not be seeing the equation of effect of gravity on various masses on the Earth.

The modern day inventors, who have the job of transforming a traditional genome may have to see beyond the apple, gravity and the earth.

A traditional goal-setter and a trainer who is assigned to manage the goal-setter will have to accommodate the three factors of Health, External Forces and the minimal Mass needed to achieve a single goal in the first place.

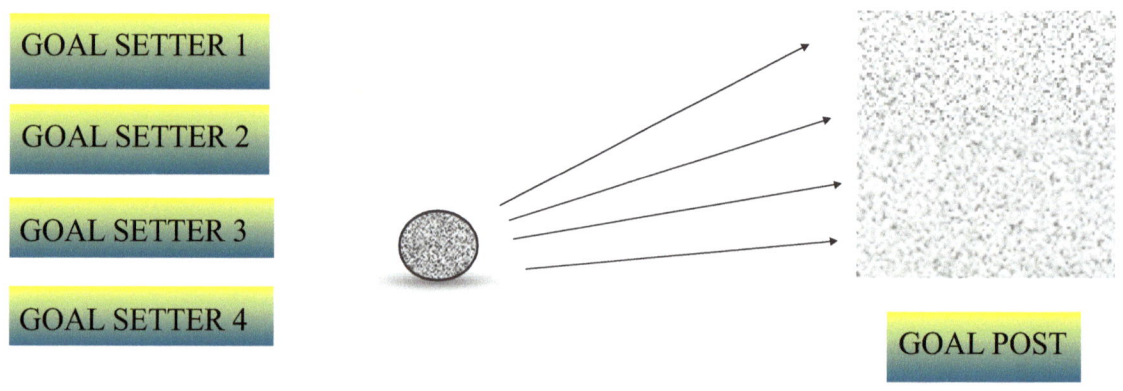

The above diagram shows a football field with four players and a goal post. In most football games, the Goal-setters other than the Goal-setter in field remain to be a Goal Keeper. This means that the set of Goal-Setters who are made to sit idle start showing a level of resistance towards the Goal-setter to move towards the goal, even if they belong to same team, same state or same country. This is already described and proved with the simple experiment of a 'Rat Race' with nothing other than Cheese kept for rats to consume in a 'Rat Maze'. The resistance in a way helps the goal setter to move forward with the path taken. The resistance, if taken various color forms then one may call it a Social Stigma. The resistance gives one the necessary potential to gather energy. This can be proved with a simple experiment of a water tank kept in a high rise building. The energy needed to pump the

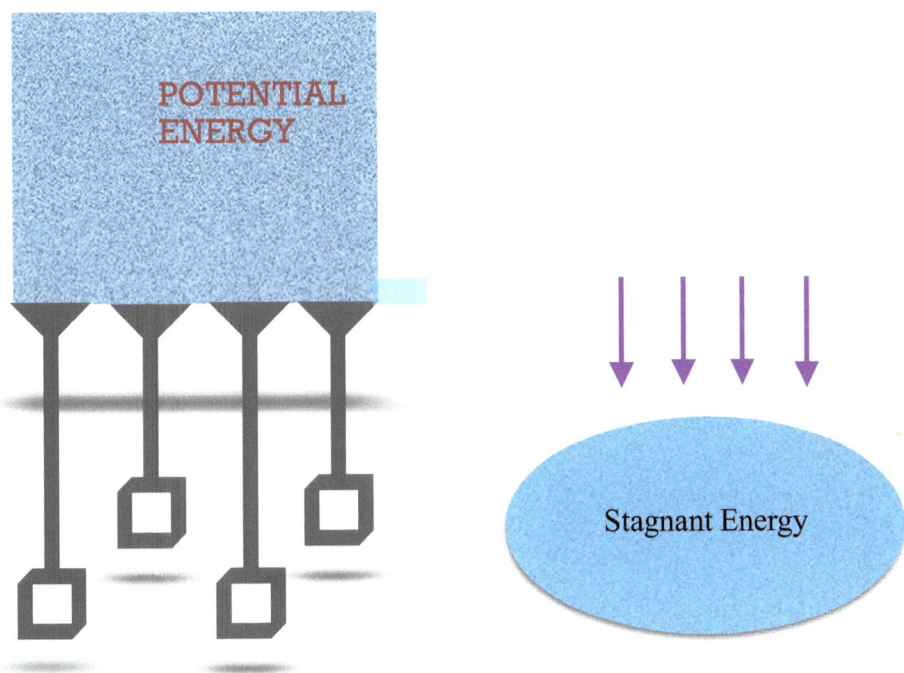

water to the tank situated in a high rise building, is gathered by the water molecules when moving up the building. The potential energy stored by the water in a high rise building when compared to the potential energy stored by the water in a pond on earth is large in value, this gets converted to kinetic energy to provide turbines, the energy to generate electricity.

If the Goal Setter, gets to show a great lead potential and also shows a good amount of resistance towards dissipation of unwanted energy towards the goal, then the other Goal Setters will follow a similar pattern in building resistance. Resistance built to a good deal of an extent, increases the endurance of the product since it reduces free radical entry into the body. Nor it allows a free radical multiplication thereby helping improve the resistor's value to a higher quantum. One has to remember the point that greater the heights one reaches by inching towards the goal, greater the resistance to be included in the daily practice so that the nervous system remains stable instead of becoming a fluctuating voltage with no resistance. In fact, the inventors decided long ago to make a bulb glow only using a particular voltage and inductive current instead of a large voltage to begin with. This clears an important point that the stagnant energy and potential energy co-exists how much

ever efforts have been incurred to increase the potential energy, the stagnant energy will act as a resistor in accordance with the Ohm's Law of physics.

The resistance acts as a checkpoint in measuring the levels of threshold for tolerance. When one is marching towards the goal, the tolerance is measured by inserting resistance at equal intervals. The previous description of resistance shown by other set of players in a game of reaching a common goal, may be termed as Zero Sum game in which there is a compensation or a penalty to be paid by other set of players for having shown resistance to the movement of the single player. The summation of resistance and the direction, magnitude in which the the resistance is shown by a player. This can be observed when two trains are crossing in the opposite direction versus two trains moving in the same direction at different speeds. The relative velocity of two trains crossing in different direction is almost double the speed of one of the trains when compared to trains moving in same direction. When two trains are moving in the same direction, the relative speed is a subtraction of the speed of one train and the other. The train may look stationary or non stationary based on the zero value versus non zero value of the relative speed.

A Bull and A Bear has a good level of similarity when compared to Movement and Resistance. In an environment in which a bullish attitude is inflicted among the greater set of masses to achieve the individual goals, and improve the nation, a bearish attitude arises either from the set of internal or external forces. One may argue this to be a healthy balance of forces, however at a point of time we will have to necessarily check the proportion of the bullish versus the bearish attitude to provide individual resistance via provisional identity. It boils down to a simple point of 'Who owns the Energy' versus 'Who owns the resistance'. Greater the resistance, lesser the excitement and higher the good for the mass on a spiritual basis. Greater the energy, lesser the provision for time to think of sidelined goals and higher the good for the mass on an economical basis. It is necessary to balance the two to provide a healthy living and thereby inching towards the goal, be it a six sigma industrial establishment or a foot ball field where the top icon player moves towards the 'Goal' on a regular basis.

The color coding of a resistor in general tells us the level of impedance it provides for a specific electric circuit. One has to be careful in handling electric circuits, since the resistors in general produce heat as a reaction of resisting electric current. For higher electric current, the resistor may simply give in to the reaction and produce more heat there by damaging the circuit. A Goal Setter, can take hints from the resistor in physics by allowing only specific levels of inputs to the nervous system and not create too much heat to the nervous system by producing more inputs which causes the nervous system to interfere to heal the specific zones of stress and there by producing heat allowing tiny organisms to take over. An idealistic goal and path to chase would be to have a minimal number of high resistance healthy goal setters around to take inspiration from and move on with the path chosen.

In a real time environment or a simulation of a foot ball game, the number of forces inside the arena are around 25 in number including the Goal keeper and the referee. The team which generally is composed of trainee kids with a majority of them having the capability to be composed and move the ball towards the goal post without much penalties from the other side may reach the consistency of winning large number of tournaments instead of being judged with a single match. The team which is on the opponent side, can get into an experiential mode to understand the attributes, analyze and implement to improve to the next level than the level in which they currently are.

'Resistance' and 'Reactance' are two specific terms of interest in analyzing the offense and defense (sports) groups in specific competing teams in a football field. Offense teams are mostly reactive in nature, may be impatient towards a long wait time and naturally have high energy potential untapped. Defense teams, however have a knack of converting themselves to offense teams at any point of time by tuning their mind. Though the characteristic of a defense team may be branded to be similar to a resistor with high levels of resistance towards the task at hand, the actual characteristic of a player will mostly tend towards offense, since any player in a football field would always be on the look out for a goal. Goal Keepers on both the sides remain to be offensive and defensive alternatively when the ball is moving towards them, deciding the future of the game.

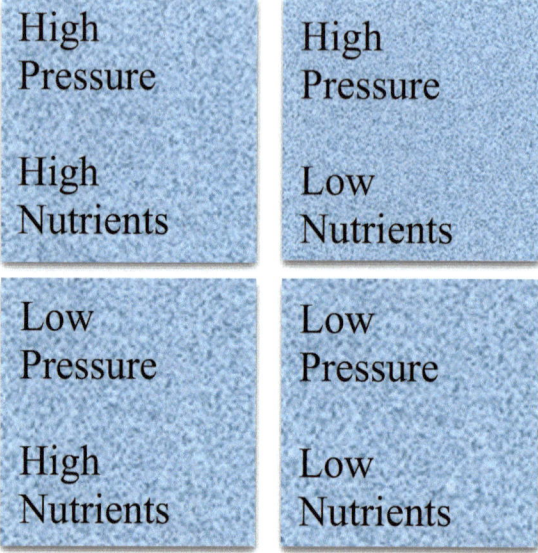

Let's assume that gravity as a force is dwindling on the Earth surface, then the pressure necessary to pump the blood would drop to the extent of reduction of gravity. Gravity as a force keeps the forces alive to pump the blood through the central nervous system and help in the process of purification by selective discharge of the plasma to specific organs. If the gravity is becoming low, the pressure to pump becomes low because of an upward pull of energy. At this point, the level of nutrients may be kept at low based on minimal push of nutrients to the necessary organs. It may be compared to a transient state or a meditative state in which a goal setter shows least resistance to the opposition or resistance created by other set of players. This comes with a level of maturity. On the other hand, the low pressure to push the plasma upward also results in a stagnant energy towards the lower part of the body shown in the previous diagram of potential energy and stagnant energy. If the stagnant energy is towards the lower part of the body, it results in possible general physical fatigue. A low pressure and high nutrients in the body may not be a fruitful combination as far as purification and cleansing goes. An upward surge is shown as a triangle

on the right side of the figure to indicate the natural pace maker's innate ability to keep deriving energy from various elements of nature to keep pushing itself to do the job.

The other set of quadrants placed in the opposite directions such as High pressure, Low nutrients and Low pressure, High Nutrients indicate the moderate feasible environment for a neonate to survive and flourish, since provision of High Pressure and High Nutrients may not be available for all the neonates in an ideal environment and will go against the Social Stigma, survival procedures. The quadrant diagram above depicts the process of eliminating a difficult atmosphere of less nutrients in a low pressure environment so the damages to the nervous system can be minimized. The process also provides a point to understand about breathing technique described by many of the masters of the arts.

It takes us back to the story of Hare and Tortoise again in which the Hare, to provide the initial momentum for the tortoise to move, has to breath heavily and in an environment without proper food and ventilation, may result in the Hare becoming tired and less immune when compared to a tortoise which moves extremely slow in these situations.

If we consider a large scale hotel environment in which there is a strenuous job of feeding masses on a daily basis, the level of nutrients to be absorbed by the cook preparing food can be at the highest level which later percolates to the lowest part of the food chain in the group, the consumer of the food. It does not matter whether the food is consumed on monetary terms or prepared for a person born with a silver spoon free of cost. It can be compared to trees having the highest absorbing capability in the roots, except few which can survive even without roots named Bryophytes.

The quadrant by all ways and means, tries to portray the stringent measures one has to adhere to in order to get to the level of slow, low pressure zone with high nutrients getting pumped on a subtle basis. This takes us to an important point of discussion around the heaviness of various nutrients and any metals (which may be produced as a by product during the reaction) gathered along with the plasma. If the nutrients needed for healing are of higher atomic mass, the pressure needed to pump for the pacemaker is high and hence takes a longer time to send the nutrients to the specific zone for healing to occur. This in turn will have its own wear and tear, which needs to be fixed apart from the zone which requested for the nutrients. One interesting query which may still come up is that if the lower part of the zone takes lesser time to request for nutrients than the upper part of the zone.

The idea of introducing an upward surge and downward drain by showing four quadrants of a Pressure-Nutrient mix would partly let us know about the ability of one's body to increase or decrease the pressure based on certain scenarios. Its good to know about fight-flight response of the body during critical conditions versus the feel good scenario during a relaxed tenure. Fight-Flight response is more like a body's reflex capability that is inherent in the genes to respond to an emergency. When the senses start perceiving a lot of information, the body tries to accommodate the rest needed to cool down the heat already produced due to the perception. This is generally through the reflex mechanism of the nervous system in action. To understand more on this subject of conscious and sub conscious mind, one has to research further on the topic to bring both under one's control to proceed with the Goal and the Vision towards one's life.

One may think about the goal as an infinitesimal set of steps instead of a one-day activity if the body produces a signal about the rest needed. Few individuals identify this within less time, introduce specific measures through external means and get to restart the activity by reducing the time spent. The athletes and artists working in their own specific fields without diversion for very long time to produce a work of art, shows that they are working on inclining their body and mind to accommodate the goal first, assimilate it on regular basis and allow the subconscious region of mind to become active and produce great results. The quadrants may also be slightly compared to the chambers of the heart, the evolution of a human heart with 4 chambers allotted for specific purposes. The 4 chambers in a way remain to be the sac of collection, pumping, replenishment and monitoring the health of an individual goal setter on a day to day basis. It is little sad to see that there are now a lot of external equipments connected to the chambers of the heart, to monitor an individual on a minute by minute basis.

On a completing note, the set of Reactance and Resistance need not be emphasized to a great extent since the law of natural forces tend to offset a highly reactive element towards stability by producing a natural resistor named wind. One can witness this with high speed winds in certain areas of Earth, due to the pressure zone created by reactive elements and by products. Resistance in a way, helps stabilize an operation to bring the pace maker to work at normal frequency. One has to measure the goal setters ability to produce results during a said period and thereby stabilize after a period of time to adhere to the Ohm's law of resistance.

8 TH STRING

Since we started this book with a theory of uncertainty, when we commenced a first step to know what these subjects of Physics, Mathematics and Chemistry can do to one's life while and after setting a goal and to the path adopted, it's imperative for me to discuss another interesting topic in the field of Physics named 'String Theory'. String theory in general talks about few of the effects we already covered, namely the effect of a humongous goal, a goal setter might produce and the force which he or she may exert on the earth. There is a particular hypothetical particle named 'Graviton' which is covered as part of the String theory which encompasses living beings beyond Earth, effect of Earth on these particles and how they are treated and received.

It would be breathtaking to know that the list of planets, stars and galaxies together form only 6% of the entire universe. The mass of the universe measured seems to be around 10^{53} Kg at least, measured by inventors with various equations and calculations. It would cause a great deal of diffidence if the thing B with a mass of 6.64 Kg described in the initial chapters, is unable to inch towards the smallest step towards the goal in the field of education, arts or sports in the space of Earth. A simple question which may pop up in a reader's mind, on how or why a telescope which is able to relay us the images of a distant planet or satellites, unable to be perceived when seen through a normal eye. Let's analyze this with different steps from the view point of an ordinary observer. There are certain set of satellites which are visible to naked eye, such as Moon and other stars during the night.

Sun is visible during day time. The eight planets described by the solar system named Mercury, Venus, Earth…Neptune are spaced at specific distances, underwent an initial momentum by an external force for a revolution and a rotation around the Sun in the solar system, a force that is not too strong that there is a collision by asteroids and comets and a force not too slow that there is a formation of a frozen set of particles not conducive for rotation and revolution. So, in our natural assumption to understand the described story in our younger age that Mercury and Venus are closer to Sun with faster revolutionary speed compared to other planets such as Jupiter, Neptune etc.. we will have to assume certain facts, question the facts and then move forward.

If we assume that we can see a natural satellite such as moon shining bright and other set of dull stars in the night, we can definitely conclude a point that Moon is closer to the Earth when compared to the stars. Amongst, the stars themselves certain set of stars look closer to Earth and twinkle stronger than other set of stars. We can assume many of them as stars and few of them as planets such as mercury visible to naked eye, since we are able to see the Sun which is farther than Mercury and Venus, provided that there is a meeting point of the planets revolving around the sun falling in line for the observer to make note of. A Telescope for instance may be the necessity for the inventors to think of, when observing many of these natural satellites, stars and the sun. A Telescope does a great job of reflecting or refracting the image it captures when pointed towards the object even though the object is at a distant location. What may be a tiny particle for the human eye to visualize, a particle as tiny as a dust may be the actual size that a human eye may perceive when these masses are seen directly.

Some of the planets when too close to the sun, may start emitting light at different wavelength, based on the interaction of Sun's radiation with the particle they contain. The process of heat and light absorption that takes place on Earth may not be the same on a planet such as Mercury. This is simply because of the particle nature of the planet and it's uniqueness with which it was created during the initial momentum by the external force. The invention of an incandescent bulb for instance would have been done by a process of passing electricity through a filament so that it gets excited and produces heat and light. It is almost next to impossible to produce light without heat element associated with it. This was in fact described by various inventors about the impossibility of production of energy, rather the possibility of only a transfer of energy from one form to another. Even in the scenario of complete conversion of an electrical energy into light energy without exciting the filament, still we can observe a wear and tear associated with these bulbs and this may not happen unless the energy is dissipated as heat and absorbed by the atmosphere.

Every star and luminary object in the galaxy is measured with a unit called light year. The light year is nothing but the time taken for light to reach the observer, wherever she is. Sun light for instance takes approximately 9 minutes to reach the Earth's surface. Considering the various theories or relativity, such as particle versus wave nature of light and the effect of relative velocity on the light movement towards Earth based on other planetary nature, it becomes incumbent on Earth to maintain the light speed as a constant. Let us assume, the speed of light as a variable. This would result in a great set of anomalies as follows:

1)   First and foremost, everything perceived would start appearing as an Augmented Reality environment for the observer instead of what others perceive as a natural vision.

2)   The observer may start seeing the future and the past without being asked for, and when this happens with a large set of masses, it results in collision.

3)   Most if not all of the humans on the Earth, would look to be magicians and racing towards the highest pedestal of magic, with few humans left to perform the spell on.

Many of the inventors, who were in solitude were perceiving the nature of science because of this specific reason. One may not find a great deal of difference between a clairvoyant and an inventor.

What a clairvoyant sees or perceives at the highest energy level, is verified and tested by an inventor either to prove it right or wrong. We may see in day to day news about special powers of a child to remember everything and anything that it reads or an extra sense of a foreteller to predict theories of previous birth and next birth, which will just showcase the particle nature of light getting manipulated either because of the gene theory and structure of DNA or because of the solitude environment in which they may have acquired these powers. In a way everything can be converged to a piezoelectric effect in electricity, in which particles gain a great level of charge by rubbing the particle on specific other particle with similar or slightly higher charge.

Minor experiments in this field would surprise us, such as a comb attracting small pieces of paper after acquiring static electricity through rubbing or a mechanical watch working without battery for lifetime, only with the acquired charge from the movement of hands. If we go even earlier during stone age, piezo electric effect may have been the first ever process to generate fire for woods to cook food by rubbing two specific stones which has the material nature of producing fire. One can deduce a point, that there is a natural force associated in every pruning process or birth process of neonates. Milky way as a galaxy which contains our solar system looks to be a tiny filament emitting light compared to an incandescent bulb excited with electricity. There are billions of galaxies yet to be explored in the universe and existence of many of the stars still remain to be a mystery.

'Graviton', a hypothetical particle termed by the inventors till date has the capability to covert itself into a vibrational form. One may presume that Graviton is single a particle existing under the vicinity of Earth's gravity. Validating the initial theory of multiple masses acquiring the characteristic of neighboring particles and assuming that **Graviton as a particle which has eyes and ears, sitting near the mass to watch every movement or the activity measured to the level of millisecond or a nano second** may just look to be a fiction in the olden days. However with recent discoveries and proposals, the theory of Graviton being a hypothetical particle to an existent particle has started acquiring traction. This can be proved with multiple experiments and rewarding results in the field of medical science.

Medical Miracle, a term used by many of the doctors who still accept the fact that there is a force beyond them to protect, and save the forbidden mass. Let's presume that the formation of a neonate was through a series of evolution and vestigial process applicable for a specific region. Let's also assume that 'Graviton' particle in that specific region gets to decide on the external climatic conditions based on the sum total of the masses and the force exerted in that region. The process of identification of the neonate's future depends on the group of Graviton particles which at a point of time should know the past of the neonate. If we really wonder how a Graviton can identify the past, the neonate we are talking about is nothing but a mass of liquid, solid and gas. Liquid is in constant movement and Solid is renovated with nutrients from the liquid needed for survival, since every part of the body is nothing but a living organism. According to String theory, when a thickened liquid such as chyle is in movement, it carries along with it a certain level of vibration.

In Music, when a string is excited it generates a sound which at different intonation and pitch produces songs. When a group of strings is excited, it produces a subtle harmony which influences the nervous system to reprogram the chyle in discussion. The vibration we are talking about now, need not be through music alone but can be due to external sounds that one hears, a powerful speech that can have a profound influence or a daily recital which can invoke a set of characteristic. For clarity sake, we have quoted the movement of chyle however in reality there may be many combinations of the nutrients in the liquid or the plasma which can create a vibration during movement. This movement becomes the key for the Graviton and many of the vibration produced by the neonate and the characteristic resemblance of a neonate with its ancestry origin. Graviton, may simply be termed as the dualistic nature of the hypothetical particle. When the hypothetical particle is controlling the neonate, the characteristic of the neonate becomes slightly different than for a neonate which controls the 'Graviton' particle.

Invention of Stringed instruments through out the evolution of humans, lets us know about the fact that there is an affinity of the subconscious

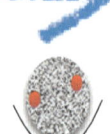

mind towards the sounds of music and beats arranged and composed together with great quality and perseverance by rehearsing the program as many times as needed. The below equipment depicted resembles an 8 stringed musical instrument which can produce a continuous vibration in the chyle or other set of circulatory liquids during the process of energy absorption through the nerve endings. When the energy absorbed is emitted through various means by a neonate, the vibration is either sent through the forms such as liquid, sound or air. Many of the sports men, get involved in a concept called conservation of energy for producing the maximum results in the tournament. In a similar fashion, a neonate with or without its knowledge and with the influence of the nearby 'Graviton' particle may get involved in the conservation of energy and dissipating only a percentage of vibration out through these 3 forms.

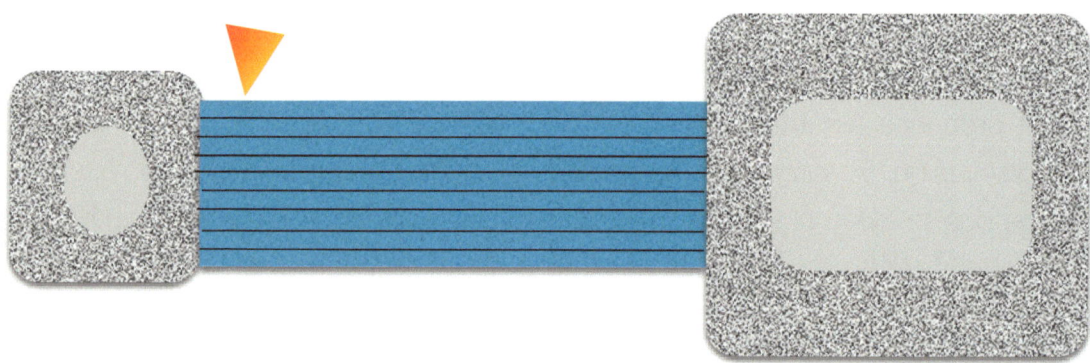

String theory, helps in understanding the nature of the universe, how milky way as a galaxy was formed and the alignment of planets in the solar system. The effect of Sun and various planets on the Earth is measured in general by the String theory and the subject of science is yet to be forayed in detail. It's also interesting to know that a baby stopping to cry when sung a song, and given proper rest. The science of subject of different sets of sounds on the body, different genres of music on the mind is a huge portion

though a minor part of it can be discussed in few pages for the reader's attention and well being, towards the achievement of one's Goal.

Sound travels at a very less pace compared to the speed of light. The reverberations and acoustics which travel through the auditory canal to reach the central nervous system and getting transmitted to the brain for perception and understanding happens for few neonates within milliseconds, few humans within seconds or minutes or hours and few neonates may be specially privileged to undergo transformation to get to that stage. One may try to experiment the behavior of sound, similar to behavior of light in various mediums such as solid, liquid and gas in order to derive an equation out of it. This may provide various clues to understand the nature of interaction of sound and light during one's perception of his or her goal. We are talking about only sound and light, since these two are the major components used during the Goal achievement process over all. Skin and Nose generally are used for vital functions such as touch sense and breathing. So usage of these two for the process of inching towards the Goal may not be advisable.

Its important to know that the effect of light on one's nervous system is no comparison to the effect of sound. A lightning and a thunder are two examples which are produced at the same time, however perceived by the observer at two different time zones. We may vaguely know that the Sun by itself is undergoing enormous nuclear fusions, which is resulting in we seeing any thing that we perceive through natural light for inching towards our Goal. We may also need to know that there are by products of the reaction because of the fusion such as Ultraviolet radiation which is harmful to a neonate's existence and hence the set of Graviton particles would have undergone transformation based on the inputs provided by the nervous system of a large group of neonates. This can be proved with a simple experiment which provokes a fear factor in a neonate when it encounters a predator. Let's presume a neonate which is a herbivore is existing gracefully, however a rising population of a carnivore is a threat to the herbivore. The herbivore in this situation cannot do anything other than undergoing a natural palpitation of the heart, there by increasing the awareness of existence. The palpitation directly results in movement of chyle based on the

sounds and light particles the neonate has already gathered in it's life till that point of time.

The direction and magnitude of movement of chyle and plasma of a neonate cannot be defined well since controlling a subconscious mind is like controlling the nature of light. The neonate in a sense may think of its form which may be a larger herbivore. In modern day theories, undefined during the previous theories of evolution what one may deduce is about the group of neonates sharing similar attributes making a slight movement towards the group of other neonates sharing the attributes with an overlap. This plays an important role in the chyle movement process. For instance, a herbivore with a specific attribute may look healthy and large, becoming a feast for a carnivore or may undergo a natural transformation to the next available biggest or smallest herbivore based on availability of natural resources. This can be illustrated through Venn diagrams with overlapping curves.

The above figure depicts the co-existences of various neonates in a forest zone. Neonates starting from herbivorous to carnivorous origin. The overlapping zone tells us about the characteristics they share such as walking with four legs, having two eyes, breathing with nostrils etc... Let's presume the red circle as a carnivorous zone which encompasses all the characteristics of the food chain such as herbivores to other carnivores. If we take a survey on the red zone versus the other set of zones in the food chain, a clear picture of very less proportion of the existence of red zone will be evident since there is a high level of competition for the prey and high cost maintenance of the offsprings. If in case the carnivorous animals start grazing green instead of hunting for prey, the nature of herbivorous zone may change since there will be a dearth of green grazing areas which would result in a competition for food, resulting in delinking of the food chain. It's also important to see that there is no existence of a consumer beyond the red zone, since the nature of the red zone is to share most of the characteristics of different set of zones starting from zone one. In a way, the red zone also tries to gather the vibrations gathered by the other zones, such as the perceptions of vision, hearing and sense. There is a problem associated with the red zone that it is more of a symbiotic type, and it becomes a prey for the disease causing infectious agents carried by the other set of zones while consumption and hence becoming a prey by itself.

The advancement of science and technology in voice transmission through various protocols, whether lossless or lossy has made a mark in the field of Energy Physics. Lets say the voice is converted into electrical signals with a microphone and made to travel a wide area network through connected wires without any breakage and reach the other side where the transmission has to be completed with a loud speaker which can receive the electrical signals, the conversion process has to happen within seconds. If it fails to do so, then there is a noticeable lag in the communication process and the invention becomes a brick. Every time a user has to see an advancement in the field of these inventions, there is also a noticeable increase in the amplitude or the frequency of the electrical signals transmitted. Beyond a particular threshold, there is no further amplitude or frequency manipulation possible with the energy physics since this may interfere with the normal existence of the other frequency waves and distort the spectrum. A natural vibration of the 8th string in one's body

and mind can help in advancement towards the goal with the path adopted as sinusoidal over the course of time of the Goal.

For a Goal setter, the capability to connect the facts gathered over time adds good value in moving towards the vision. If the facts presented are absolutely disjoint, it would make sense to store the facts and connect them later to provide a vivid picture of the future vision instead of getting to a conclusion at the first sight. The provision of natural and artificial measuring scales such as Virtual Reality tools and Graviton particles would help in showcasing the epitomes in the field who have already spent a great deal of energy in completing their works successfully with flying colors.

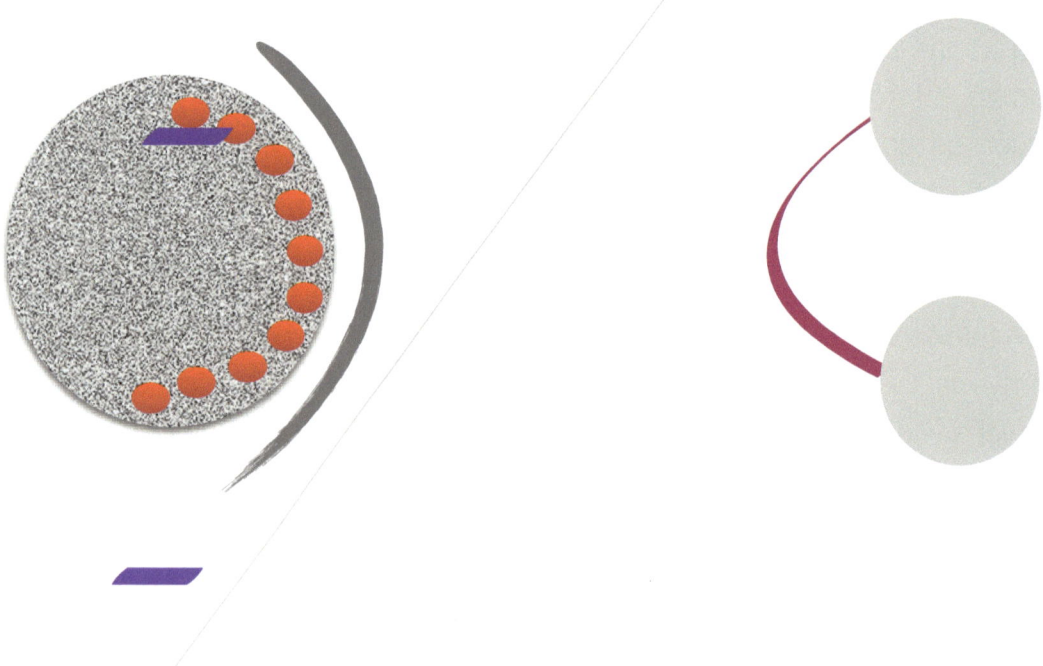

SURROGATE KEY

It's important to bring to one's knowledge that the world has undergone an enormous amount of refinement in every field. Be it, Name it or Verify it, there will be a great transformation already in progress or at a threshold. Let's for instance check the field of Telecom. The information which took so long to get transmitted previously is now getting transmitted within seconds. The satellites which are orbiting around the Earth have resulted in making this happen. It's also heartening to know that there are options for users to receive every product and every service at doorstep. The most advanced latest addition to this is an intriguing concept of electronic commerce. How, a coin or a currency gets converted to an electronic signal and gets transmitted to a large distance like what just happened to sound and light in the previous passages and now reconstructing itself at the receiver's side. The Telecommunication is not only about printed circuit boards with resistors, but also about the design of the package in total. What we failed to mention in the passage of resistance is about a zero resistance theory and superconductivity. It's a Quantum phenomenon, which makes the current passing through a material to be infinite when the resistance becomes zero in a conductor. Though this is technically an impossible condition in most electronic circuit designs, there are special cases in which a magnetic levitation or repulsion takes place when an infinite current passes through the material without any resistance. We can see this through a trivial example of a magnet attracting or producing a repulsive force on another magnet.

The material property of a superconductor is to allow infinite current with zero resistance at a particular threshold of temperature.

The emission of magnetic flux at a point from the conductor results in a change in the behavior of the conductor bending towards the property of a superconductor because of the electromagnetic property.

Information in today's world passed through conductors at a point of time may need infinite electricity to pass through the conductor for attaining the speed with zero resistance and may result in conductors becoming superconductors finally in levitation of the materials.

Levitation, an art generally performed in magic shows by illusionists is now made possible in physics with proof. The incubation of this idea would have made origin during the time of discovery of the land mass when navigators used compass to check the direction towards which they had to use to travel in the boats. A Magnet in general has a polarity of showing us a north direction when allowed to hang freely. The Earth by itself is an example of a superconductor, in which the rotational speed produces a magnetic polarity in the edges and considering other set of planets producing similar polarities, it will result in levitation in the path that it has taken. Superconductivity is a concept which moves a huge train at high speed. The enormous potential and kinetic energy to keep the train levitated and move at a very high speed is to be seen with an extra lens. It's almost like an airplane levitated without wings. It is important to know that the space between the elevated train and the tracks is to be left undisturbed when the train is in movement before getting to a halt which is achieved by modifying the electromagnetic effect on the tracks.

Maglev, though the concept is much used in many of the applications such as automotive through physics, it can be applied to a Goal-Setter's life cycle too. It would take us to some of the previous quotes which illustrate about the effect of a rush of plasma to the organs at work when you are exerting yourself a great force to move towards the Goal. The plasma which we are talking about may be through the multitudes of veins, arteries, capillaries and the pressure or force in pumping them to the necessary organs depends on the various aspects of angular momentum, positional velocity etc... It can be compared to a car moving at a very top speed and still able to make a turn in a road which is having bends, without applying brakes. This can only be possible if the driver knows the level of acceleration to maintain, the radius of the curvature and the position of the different kinds mass inside the car and its effect on the acceleration.

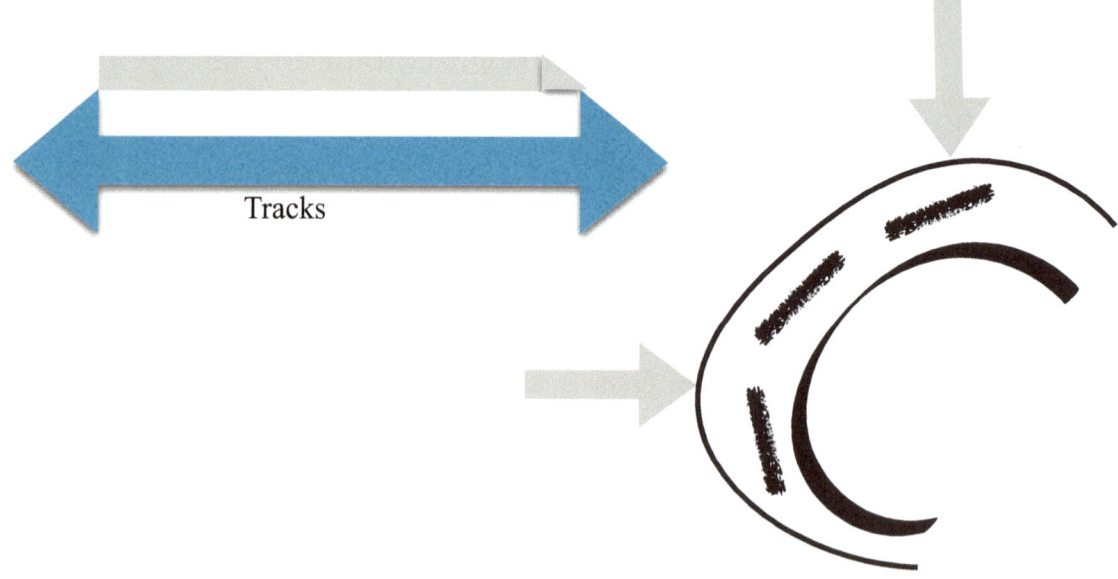

Tracks

What happens in a Maglev in the tracks by applying electromagnetic effect on the train, happens on a car without any external electricity applied but by a mere force of acceleration and an angular torque, giving the car a slight drift in order to balance the other side of the car to stabilize and in some cases, an elevation. Elevation also happens when the car is made to ride on a bridge downwards after climbing at a great speed, in a Linear road without any angular momentum. Elevation is about exerting a nominal force in the path for a goal setter and bringing it to a halt at regular intervals. Hopping at regular intervals with an elevation may prove to be an infinitesimal step instead of a 'Maglev' levitation, for goal setters who are extremely focussed and rigid in marching in the path towards the goal already taken. We can question this theory using an example of how a wood floats on water however large it is while a small piece of iron sinks, but a Ship of a huge size floats while pedaled using turbine motors on an ocean.

The material property of a wood is to float on water because of the buoyant nature of water and atomic nature of the wood. Wood being a natural resistor in insulating electricity, also exhibits the characteristic nature of floatation in water which is even now widely used in many fields such as construction and circuit boards. A ship which is made to carry a large number of people may have to be designed with great care to resist the buoyant force of water when moved with the right displacement speed. The fuel and the energy provision for the engine will be decided based on the mass inside the ship, the nature of materials used in the ship and also the seating arrangements. This also depends on an important factor to be considered, named the climatic conditions and wind speed. There are scenarios in which the most astute design may fail because of the natural conditions not suited to nurture or ignite the movement. The capability of the ship to hold the travelers, of a maglev train to move the mass or an airplane to transport designated executives from one location to another without any difficulty and a good level of nurturing and feeding may be termed as Surrogacy in crude terms.

Surrogacy in a modern day environment may refer to a women's capability to deliver a baby with a legal approval and financial support from partners who would want to bear and rear up a child. Surrogate key in technology means an individual index in a table of data which would provide a reference key to another set of indexes to search a term. This helps us with various sets of applications such as banking, commerce and trading to account for day to day transactions, since the number of transactions in a bank or a top retail store may be in millions. In case of an anomaly, it becomes important to retain some of the reference keys which also includes a surrogate key to deduce and point to the right data without losing the integrity.

Its quite interesting to know about the evolution of a neonate to hold a baby inside the body from a premature concept of egg laying mammals in which the egg is allowed to hatch and we see a living creature out of it over the hatching period. The holding period differs between different types of mammals and within the same mammals, different types of eggs. This would necessitate about the mentioning of the novel invention of DNA (Deoxy-RiboNucleic Acid) and RNA (Ribo-Nucleic Acid). The DNA determines the characteristic of a neonate during the formation process. DNA and RNA both are part of a gene of a neonate called chromosome. The chromosomes are part of every cell of an organism, including a unicellular organism. This includes the skin cells, nails and hair of a neonate. In a real time environment, any neonate formed from a surrogate mother is more of a fusion of triplet cells to form the neonate by birth, than compared to an environment in which neonates are bred with care and devotion, from two parent donors. The entire process of formation of an embryo, with umbilical cords feeding the neonate from the time of formation till the delivery may be in crude terms called as surrogacy in which the organs of mother start feeding the child through an instruction from the nervous system which may not be known to a surrogate mother.

With the advent of technologies and environment, it becomes imperative to accept the reality of rapid industrialization and the profound effect it has on living neonate mothers, it is also important to verify the fact that the neonates with the capability to produce eggs are protected with extreme care in developing nations. The negative effect of the legal battles at a point of time for monetary or novelty terms, for a living neonate during later years may be negligible when compared to the capability of a neonate to produce healthy eggs and result in strong offsprings. The effect of modern day technology in this entire process of a surrogate offspring is intense since these technologies act as a platform for the nurturing environment, and this may not have been possible otherwise. The process of surrogacy plays a key role in all other processes since this decides the intervention of a human versus humanoid in the future. The subtle difference between the two would slightly differ in the way in which other humans and humanoids are handled by the other humanoids and humans through a conceived logic and preprogrammed notions. The modern day humanoids have the capability to perform every artificially acquired intelligence and logic that are similar to a human. The proportion of humanoid and human interaction makes a difference in the production of both.

Apoptosis or programmed cell death is a process in which the cells undergo aging and shedding in a natural process. This can also be referred to as an elimination of natural waste products which become a byproduct to the chemical reaction of heat during digestion. These by products are acted upon by millions or billions of bacteria in the gut and the intestines to enable a symbiotic relationship and at a point of time allows only a part of the byproducts formed to get to the final stage of elimination in the large intestine, colon and caecum. The entry of beneficial bacterial colony in a neonates digestive system is a wonder which makes one think of it as a medicine. There are other options such as live culture of these friendly bacteria which can be swallowed as tablets or even mixed with tonics to improve the process, however the natural way of enabling it to grow in the intestine and the digestive system by using the venous and the circulatory system wisely seems to be a viable condition in today's crowded scenario. A surrogate's ability to nurture or hold a living neonate inside her can be enabled mainly with favorable atmosphere in which the uterine bag is able to identify the newly formed neonate as a foreign body, and still hold the body inside the bag without any discomfort such as deriving food from the neonate instead of feeding food.

Though the crossing over and formation of a unique gene for the neonate is in an external environment with the support of external technologies, the capability of a surrogate to hold the neonate for the gestation period and feed it with every beat of the heart through an umbilical vein is already proved to be a nature's work in action. Any external interferences caused to the surrogate through supplements, may only work to be a provision for storage of food to feed the newly formed neonate post the initial formation of body parts similar to the surrogate, with the genetic code derived from the nucleotides. If the genetic code derived from the donors seem to match well and produce a neonate during the initial phases of the gestation, it grows with minimal support needed from the surrogate. However, if the genetic code derived from the donors have a deficient protein or specific components which is needed for the neonate's growing ability inside the surrogate may derive the needed energy and components from the surrogate, if unable to do so may derive from the external world elements such as light, sound and other forms of energy.

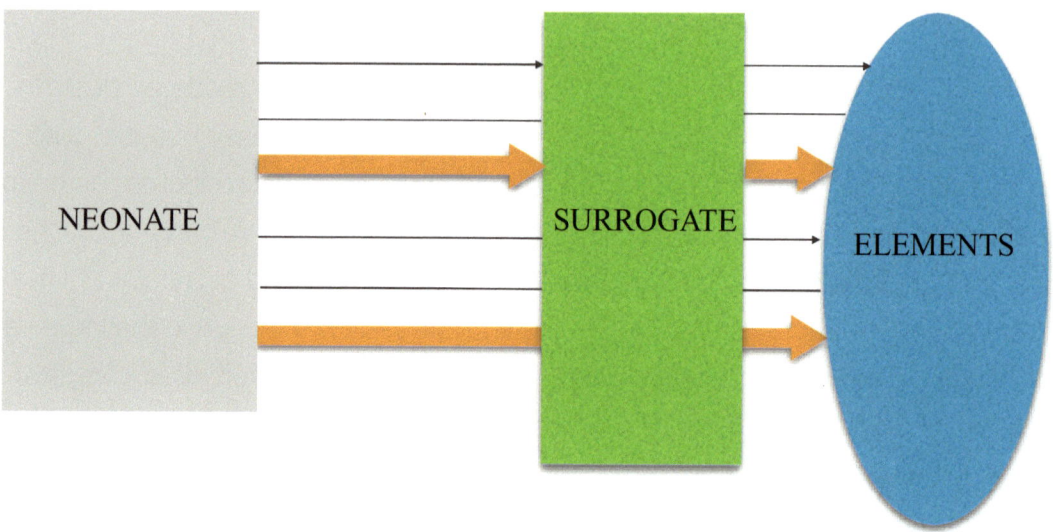

The above figure depicts the connection of a neonate with the surrogate and the connection of the surrogate to the elements. In many of the successful surrogate processes, the reaction is successful since the surrogate is successful in fetching the needed elements from the Elements zone. In a way, we can deduce that the surrogate is successful in discerning the information of what is right for the neonate and what is not. Hence both the surrogate and neonate are able to live and flourish in a symbiotic environment. In a real time environment, this may or may not be true. The characteristics which the neonate derives from the parents may take a back seat and the nutrients may take a front seat, and at a point of time the neonate may become the offspring of the surrogate through the establishment of the food source through the umbilical vein.

One may wonder on the inception of a first formed neonate without a conception and born not through a surrogate. This initially was through a process called fission in which the eukaryotes used to divide themselves into two, four and eight. This was primarily the way for the microorganisms to reproduce and flourish in which the surrogate splits into two and the part that is split becomes a new surrogate. One may witness the microorganisms which may not be so beneficial for the elimination of waste products from the body, rather divide and proliferate in large numbers near ponds and lakes which are stagnant. It's natural for these organisms to co-exist and undergo evolution to modify themselves in different forms through the same process we talked previously. The nucleotides and the arrangement of sequences in these colonies is almost similar to a surrogate giving birth to a huge number of small neonates without an external bag and an umbilical vein.

A Goal Setter's capability to be a Surrogate when needed and to be a neonate when asked for will ideally measure the attributes towards successful completion of the small tasks at hand. In a corporate environment in which large tasks are accomplished such as maintaining the data records of a billion people and establishing each of them, their needed identities is not a small task. This can be accomplished only by employing at least thirty percent of the billion people for whom the activities are to be performed to inch towards the goal. There may be violations, integrity issues, lethargy and provisional change in the direction of the Goal when there are changing priorities from different nations with different sets of people having connections with the set of people who are involved in establishing identity for the nation they are working for. However, this can be fixed with disciplined training, planned rewards and graceful marching towards the daily activities at hand. There are various techniques of the trade such as six sigma, project management and quality assurance used in many of the corporate companies in order to assure about the product that is being developed and to be delivered in the future.

A Surrogate on the other hand, may not be able to provide or adapt to many of these techniques since all a surrogate needs is an environment to digest the food and feed to the neonate during gestation and after gestation. Neonates may not ask for nutrients directly since many of the parts to request for the specific elements would be still in progress on development, however the neonate may show symptoms by eliciting the energy via surrogate or directly from the external environment. The second choice may prove to be an immaculate conception, a medical miracle termed by the doctors since the nucleotides from which the neonate has been formed was out of strong programming on certain attributes which needs it grow, however the surrogate may just be a provisional support for feeding the neonate. In general immaculate conceptions are rare and has to be identified at an early stage to make sure that the surrogate is able to provide the provisional support. The Path for a Goal setter in case of taking the role of a team player may just be to help in providing initial momentum towards the goal of the Surrogate to just be a surrogate. The technology and environment could be kept as enablers in this process.

The Surrogacy process and a Surrogate key holds the future of a Goal-Setter and her or his capability. The accuracy of genetic coding during the initial formation process and the nucleotide information impregnated in a Surrogate's womb determines the neonate's ability to leap to greater heights without a difficult stretch and healthy steps. Alterations in the nucleotide information during cross over or a specific production of a nucleotide combination in a laboratory condition will result in a neonate tending towards a humanoid characteristic. This would be viable only if the industrialization has exceeded the tolerable limits and the need for a manual labor in factories and establishments has a very high demand for a sharp attendance. Surrogate key on the other hand in today's information technology world holds the identity of a human, attributing various characteristics gained during the lifetime from various sources through the different senses.

The maintenance of both the surrogate and a neonate in a world with rapid industrial development in order to automate many of the processes and maintain a healthy balance sheet and a moderate profit and loss statement has a bleak outlook in terms of incurring costs. If the surrogate is used for a purpose of industrial development, the effects and output of the surrogate will be different when compared to the output produced for a non-profit motive of feeding food and home making. In all ways and means, the surrogate and neonate will always have a hypothetical 'Graviton' group of particles watching over them whether it's an office premise or a home atmosphere. Surrogate and Neonate are measured for the output they produce and the efficiency of production. Much of the output may be intangible, may not be seen through eyes however, it will have an effect overall in the development process of both, post gestation, growth and adult phases.

The mitotic and meiotic cell division can be held as the root cause for many of the abnormal cell divisions which is inherited from the external atmosphere, food that is ingested and the chosen that is digested. If the body decides to digest much of the food that has a characteristic of meiosis, then the food chosen and digested would be having similar characteristics and so would be the cell division with the nucleotide information encoded. If the food and the atmosphere is more of tending towards the mitosis and the food chosen and digested is of the same characteristic, then the newly formed neonate will have very high resemblance as the parent cells deriving the characteristic from the nucleus. All these can be compared to the initial phase of the formation of the galaxy and the planets which tend to have an atomic structure with revolving electrons around the nucleus containing protons and neutrons.

The impact of a Surrogate in the process of care-taking of a Goal-Setter has a profound influence in terms of encoding the genetic information in the initial stage, providing initial momentum when making a decision to pursue a goal and finalizing a position in which the goal setter needs to land thereby making the goal setter's journey as a complete package in order to inspire the other players in or outside the field. The levitation concept discussed may just happen to any of the Goal setter with continuous and relentless efforts with utmost devotion. The Elements Zone from which both the Surrogate and the Neonate derive their characteristics may take up a proportionate percentage of influence in making of the Goal Setter and the pursuit of the path decided for him or her in their lifetime.

ENCEPHALON

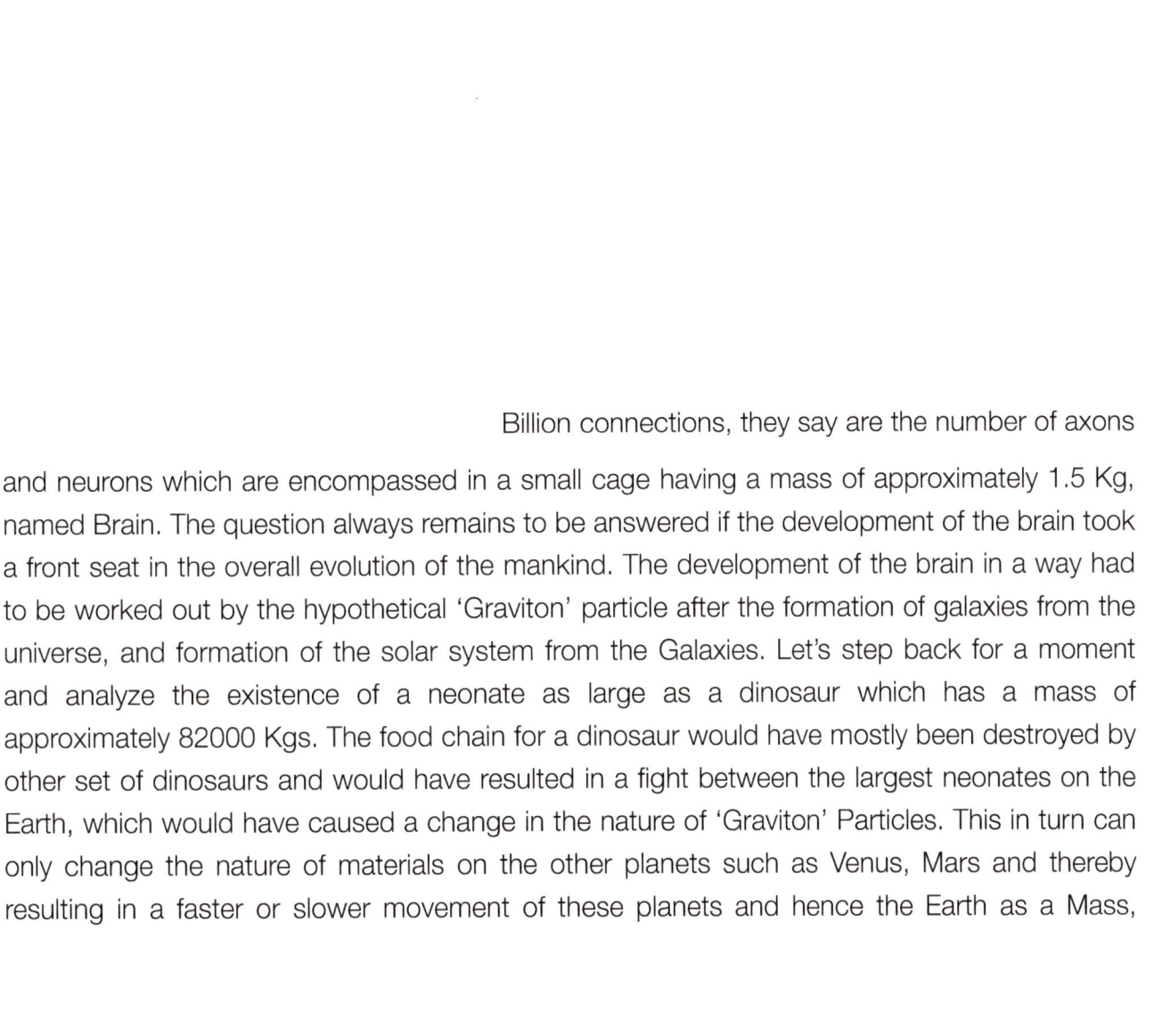

Billion connections, they say are the number of axons and neurons which are encompassed in a small cage having a mass of approximately 1.5 Kg, named Brain. The question always remains to be answered if the development of the brain took a front seat in the overall evolution of the mankind. The development of the brain in a way had to be worked out by the hypothetical 'Graviton' particle after the formation of galaxies from the universe, and formation of the solar system from the Galaxies. Let's step back for a moment and analyze the existence of a neonate as large as a dinosaur which has a mass of approximately 82000 Kgs. The food chain for a dinosaur would have mostly been destroyed by other set of dinosaurs and would have resulted in a fight between the largest neonates on the Earth, which would have caused a change in the nature of 'Graviton' Particles. This in turn can only change the nature of materials on the other planets such as Venus, Mars and thereby resulting in a faster or slower movement of these planets and hence the Earth as a Mass, attracting a comet or an asteroid to cause extinction of the race. The theories suggested by the inventors can only be ascertained because of the existence of large animals which still resemble dinosaurs, such as African Elephants which have a mass of approximately 3000 Kgs. We can even assume that these elephants have a reminiscence of the dinosaurs which existed millions of years ago, and now eliciting the attributes from a surrogate mother and other attributes from the Elements Zone, to let us know about their existence long time ago.

The brain and its function in one's Goal has already been illustrated in detail, however it would be good to validate the effect of an extraordinary focus on the task at hand and how the senses undergo an evolution when Goal Setter is in the path of walking towards the objective. Most Goal setters in the initial stage look for the next step in their daily activities. The next step may be a simple one such as giving enough rest to the senses so that they do not undergo too much evolution nor, they stay behind too much. One may presume the point that every part in the body has grown from a small organ with a mass of 1.5 Kgs, named brain in a neonate. Continuing from the previous discussion of a surrogate and a neonate formation after a gestation period, it's difficult to predict to high levels of accuracy about what is formed first for a neonate. Through many of the scans at different intervals, we can understand that specific organs of a neonate are formed during different timezones such as hands, limbs apart from the other set of vital organs for an independent functioning inside the surrogate's womb.

One needs to appreciate the fact that, the neonate's ability to grow directly depends on the Surrogate's ability to be stable and understand the effect of various forces on the body and mind. The subtle force which gets added to the surrogate on a daily basis to feed the neonate counts to be a key factor on why and how the neonate gets to retain the genetic code of the donor parents. All that was discussed till now, about the Goal-Setter and the path to follow apart from the various theories in physics such as uncertainty, string theories and recursive probability to reach one's goal is applicable with a distant view and controlled monitoring of a surrogate mother and a father on the neonate. Term neonate is used for a goal-setter now, since every living organism may have it's own Goal. A goal for a neonate such as dolphin may be to perform variety arts for the viewers in a theme park. A goal for an elephant in a circus may be to perform cycling for viewers pleasure. It become hard to train the nervous system of these neonates, in order to achieve the path. However, with advanced neonates the goal is almost complete when they are born which may sound startling to the readers however is backed by enough scientific proofs. The advancement each of the neonate undergoes when it travels a long distance towards enlightenment gets added to the database inside the brain and is stored in the subconscious nature which is mixed with the 5 elements such as wind, water, fire, solid and ether. The connection and transmission speeds are extreme in nature and cannot be measured even with the advanced transmission systems at a 100 percent accuracy levels.

The caged mass is generally enriched with fluids which protect it during duress. The brain as an organ has also undergone evolution to such an extent that it is able to signal all the glands in the body to elicit nutrients needed for survival, enable thinking and cognitive ability and also provision to make way for identifying what a neonate can eat and digest. Many of the lobes in a neonate's brain are identified and segregated for specific functions such as thinking creative versus reflex activity. From a goal of balancing the food chain for a neonate to an achievement of a human in his or her own field of education or sports, in today's environment is nothing less than a nature's work without flaws. Any aberrations in this process such as calamities, large scale devastations and destructions are nothing but the work of infestation in the default process of protection of nature. The brain as a mass which was initially used in the process of communication to the body parts when allowed to be in solitude, is now able to manage communications across the world. The world wide web and telecommunications field allows one to communicate to the others sitting across the world in a different timezone with the knowledge of a different language. The ability to perceive, understand and communicate within seconds have been tested for humans to produce arts of great magnitude. For instance, a Goal of a Pilot in an airplane will be tested for great levels of reflexes, persistent attitude to not give up during extraneous conditions and build the strongest level of confidence before, while and after the take off and landing.

Graviton, as a set of particles may act as a protective or destructive force based on the level and nature of thinking. This can be proved by taking an example of a mass healing ability of certain set of people not with extraordinary or magical abilities but with high level of confidence in what they want to complete. Its interesting to know that the Frequency Modulator Radio as an invention is able to receive signals in the air which no one can see or hear when at ease, from the transmitter's side. For instance, there are so many things getting transmitted in the air, that there is no dearth for information in any nation. In a calm mind, which is as clear as a running water in a river, the nutrient flow to the brain and back to the systems is taken care of without much strain and healing is faster and better. However with the mind clogged with many of the inputs from a real world scenario, there starts a healing crisis in which the mind refuses to co-operate with the other parts of the body with the connections through arteries and capillaries. The plasma when flowing upward or downward in the body carries a level of electric impulse which may be measured using Electro encephalogram through the brain signals.

Similar to the activity of the body which absorbs energy through oxidation and allows free movement, every thought in the mind too should be considered an activity. The main difference between the mind's activity and the body's activity is of a tangible and intangible force produced for completion of the activity. Tangible force is a body's ability to accomplish a work which is needed for inching towards the goal. Intangible force is a mind's ability to accomplish a thought process to complete the work. Going by this simple logic, the brain's evolution may have begun first before the body's organ formation. The brain can be described as a connection point of various parts of the body for an individual with a self-centric goal. The same brain can be described as a connection point of various brains of a group with a common goal. A Leader in general absorbs the positive and negative thinking of a group's energy while setting a goal. This is the first step towards the measurement of success for any organization. A Leader's ability to convince the difficult thinkers to follow the steadfast path has transformed many organizations and hence forth many nations. The leader may exhibit few characteristics unique to her, and may expect few favors in return. This is a normal way to co-exist in an ecosystem. A Leader cannot be treated as an alien with extraordinary powers to transform a society with an ordinary look or with a whip. If that's the way, the process may skew towards an autocratic rule which existed during the medieval times of world wars.

One needs to know the difference between an osmosis absorption which takes place through the skin of the body and a telepathic absorption which happens through concentration of the mind. In an Osmosis absorption, consideration is more about a dehydration of the body in which the body on the whole absorbs the necessary water to maintain the fluid balance. Fortunately or unfortunately, the fluid absorbed during the need, may come from a good source or a bad source. This would result in survival of a neonate. If the brain is advanced and the body has enough stamina to absorb the necessary fluid to survive and make the fluid get converted to the specific gene encoding of the neonate, it results in evolution towards a higher form. This may not be possible if the fluid need is continuous and dehydration takes place very often in which case the body starts absorbing anything and everything it considers as a fluid. Unless there is a discerning ability that comes by innate nature, or through a telepathic absorption of the fluid, the absorption mostly results in anomaly and an identity crisis. In a telepathic absorption, a human's mind is evolved to such an extent that the focus results in complete absorption of the content that's been discoursed. The mind energy focused on oneness results in body parts resting for that time of absorption, resulting in less heat produced in these parts and a completion of renewal of cells. There is a difference between the state of the body while sleeping and while reading. During sleep, the concentration of the energy is not at the highest level and a telepathic absorption effect can be mitigated with interference of other energies and hence resulting in the body's ability to wake up itself in the morning. However, with higher evolution a telepathic absorption automatically takes place in the mind since the mind is trained to concentrate on the oneness, anything that it thinks is the object of concentration at a point of time and hence resulting in a rest of the body parts. This may be considered as the easiest and highest form of reading.

Imagine a chid sitting in a class, with eyes wide open and doing only listening. Without absorbing anything into the skin, if the child is able to load everything into the subconscious mind only through sound and vision and thereby reproduce this during the exam, or in the leisure time without any difficulty, it results in a telepathic absorption of the content of a book like a transcriptor. **Awakening a sub-conscious mind is more like diving deep into an ocean to research the subject of the vast family of fishes.** Many of the inventors generally switch on the sub-conscious mind when they know that they are ready to switch it on. Both the process of telepathic absorption and osmosis absorption happens in the mind and body without one's knowledge. The level at which or the rate of change differs between different types of neonates and different types of humans. There are certain abilities which many of the neonates possess which humans may lack. For instance, few of the neonates have the highest ability to smell and few of them have the highest ability to see. This is a natural process of evolution which takes place in these neonates for an adaptation to reach their daily goal of feeding and getting fed. **A proper human conception takes place when a brain of a newly formed fetus, surrounded with fluids fed from a surrogate are allowed to be absorbed by the fetus in an extraordinarily brilliant but an extremely slow pace.**

The below diagram gives an input on how a brain is designed during a human conception. The small grooves shown in the brain's external covering, provide the brain to accommodate and absorb electrical inputs through a network of connections with a single sole function of enriching it with the right set of fluids and also the information to be transported from different parts of the body.

Brain here needs to be considered as two separate entities. One as an entity which acts as an organ in the human evolution which can still undergo wear and tear. Two as a transcriptor, which provides the humans, a capability to differentiate right and wrong and help guide the younger ones with a sense of protectionism without affecting other races and neonates. The brain as an organ has to be kept in a very healthy environment in order for the second to take place. This would mean that the human surrogate or a new born needs to be taken care of with extra provisions in order for them to survive the odds of the natural calamities and human interventions.

The capability of a brain in a psychic sense is same for all living beings. Right from an ant to dolphin, the measurement of the ability of a brain is dictated more by the fluid nature of the body and less by the individualistic thinking. The main difference may be the physicality in the formation of the brain itself which dictates the way in which the nutrients flow to the brain and protect them from starving scenarios. The brain as an organ, when fed with enormous input for achieving any activity will result in undergoing an osmosis absorption. This is because of the drying nature of the fluids at increased temperature for completion of many of the activities, a heat is produced in the cranial space which needs to be compensated with fluid absorption from external bodies. This only proves to tell us that an increased frequency or an amplitude in the level of thinking may disrupt an otherwise normal behavior of a neonate. We can see also notice that the brain is just an organ in many of the neonates instead of a psychic medium because of the lack of thought process generation, in terms of the extra sensory perception or a telepathic absorption. The lower forms of neonates mainly think of a food source and will be unable to differentiate when it comes to a point of survival. The EEG, abbreviated for Electro Encephalograph gives us a pointer towards the field of understanding the brain science. In this, the electrical activity produced by the brain is measured through small electrodes placed over the face at focal points. It is also good to note that the other parts of the body too produce electricity through piezoelectric effect or through movement of the plasma, conversion of heat to electricity and vice versa. However, the capability of the brain as an organ to take part in the production of electricity, lets us know about the mammoth potential it holds when allowed to work on a free will environment.

The brain as a caged organ has enormous potential to make anything happen whatever one wishes, under certain conditioned scenarios. It can simply be compared to taking care of a pregnant surrogate and the importance given to the donor parents on the whole. The fetus can only derive energy from the surrogate as long as the surrogate has enough energy that is needed for the fetus. Let's say the energy derived is an equation of a combination of both Potential and Kinetic energy at a point of time in one's life, this would simply be stated as Ep and Ek with Ep representing Potential Energy and Ek representing Kinetic Energy. Ep would be a multiplication factor of mass, height and gravity where as Ek would be a multiplication factor of half of mass and square of speed. Let's presume that the mass thing B, of 6.64Kg described initially during the introduction passages traveling at a speed of 100 Km/h would gather a Kinetic Energy Ek of 2 KiloJoules. This energy takes specific time to dissipate as heat, light or electrical energy to convert itself, with by products such as various elements of nature. The Potential Energy Ep by being stagnant at a height of say 100 meters would result in addition of Potential Energy, to the remaining kinetic Energy, an approximately 3 Kilo Joules. The Potential Energy and Kinetic Energy in order to dissipate itself in various forms can be verified with the levels of EEG results and the overall energy gained in versus energy lost from the body. The various ways to dissipate Energy from a mass carrying great level of energy is through piezoelectric technique, telepathic transformation and through sound, heat and light.

This holds an important premise in validating various theories such as Tele transportation, Mass Healing, Extra Sensory Perception and understanding the nature of different planets and their effect on living beings. The controlled dissipation of the energy can be a feasible way to avoid any aberrations caused due to the effect of high energy particles which are charged up while reaching their goals. This can be analyzed by changing the amplitude or frequency of the energy that is getting released if in the form of sound. A high amplitude and high frequency needs to be avoided and instead a low amplitude with moderate frequency can help in better results. This would also be applicable in terms of light, in which a right spectrum of colors are to be identified to avoid a group of high energy particles having a negative effect on the group of Graviton particles. The Graviton particles have an extra sense to identify and get absorbed in the host which was the root cause of creating them. Even if they are at a very large distance, this holds true. Visibility of Asteroids and comets are few examples of the effect of group of Graviton particles which can attract gigantic rocks to specific locations on the Earth and not the others. Many of the catastrophes, flood prone regions and drought parched areas are other few examples of the effect of Graviton as a set of particles.

Every particle of energy released can be in the form of a subtle vibration, or a group of aligned frequencies or a higher form of a beam. When the vibration of sounds, is continuous we may term it a musical arrangement which will have a calming effect in the mind and hence relaxing the muscles in the body. If the vibration is more of a targeted beam of high energy particles, it may find its applications in medicine such as resonance imaging and blockage identification. If the vibration is subtle, it will inculcate the characteristic of a mass healer, through out the body and mind of the producer. A Goal Setter and the Surrogates can identify these in the initial stages to decide on the path that is to be taken in the future to march towards the Goal. In modern days, a Goal-Setter is one who is only a Name-Setter whose identity is out on many of the televised platforms and the neonates and surrogates who are with the Goal setter forever, carving the shape of many of the activities on a daily basis prove to be the real Goal-Setters. The brain as an organ has to be taken care of at every stage of life for advanced humans who interact and control the advanced machines. For instance, the thinking frequency of a pilot during duress must be at a very high pace when compared to the thinking frequency of a banking cashier. The thinking frequency generally decides on the efficiency of an advanced machine, efficiency of the pilot and also of the manufacturing group. Higher the thinking frequency of the pilots, greater is the efficiency of the system on the whole. One may also undertake a group of machines to test the efficiency of a large machine which cannot be tested using a single brain but with high frequency under standard conditions.

One may worry about the increasing dependency of a set of small machines to produce a large machine. Though, the smaller machines may help in the overall development of a human and give enough rest to bring the vibrations to a subtler form, they may also bring a sense of lethargy and increased osmotic absorption of the fluid loss from the body towards the gravity. The osmotic absorption of fluid loss from the body is not only applicable for brain and cranial systems but also all the organs related the circulatory system. The frequency and the amplitude of thinking directly has an effect on the velocity and the mass of a body thereby increasing or decreasing the kinetic and potential energy of a mass. This concept is applicable on the set of small machines working hard to produce a large machine. Many of the factory produced and assembled products have a velocity and potential associated with them during the initial process. Unless, a human intervention is made possible to control a specific product which may be highly complex to be operated, by allowing the human to spend a large amount of time with the machine so that the reflex of the human is programmed for difficult conditions of operations, the machine may prove to be an overhead which can be used only for testing purpose in a factory environment.

The Goal and the Path Taken by a Goal-Setter may have implications in the daily routine, health and well being of the Goal-Setter and the surrogates as well. This can be fixed with a strong alignment of the surrogates, neighbors and the community to allow the Goal Setter to move with a good level of resistance which can be an inspiration to the Goal Setter instead of an impediment. The advancements in technology space through various inventions has helped so many Goal-Setters to move in their fields, climbing up the ladder at ease. When Stepping on a moon and analyzing the material nature of the satellite is made possible for a human, it is possible to achieve the goal set by following the stead fast path, moving slowly and steadily, allowing to gather inputs from the best possible companions and inching or hopping towards the goal like a Hare or Tortoise. The end result will show up in the field chosen by the Goal Setter, since the other set of Goal setters who have already reached the pinnacle will for sure give a helping hand to the new entry. The Education system and the training materials will provide the foundation to communicate what comes to the brain without any difficulty and hence using any obstacles faced as a stepping stone if there is no ladder.

Training of brain and mind is like taming a Wolf and a Lion. The higher the force, the larger the force with which it howls and roars.

There is nothing wrong in rewarding oneself for what is needed to be achieved in one's life unless the reward system proves to be a threat.

Activities in the right direction and magnitude, produces synergy at the right time needed. In a Modern day environment, achievement of a great goal may become a common activity.

Achieving a great goal, with right synergy within the stipulated time with right direction and magnitude may decide the course of action between surviving companies and flourishing organizations.

Education and the system to implement, on the whole must have adequate amenities to improve telepathic absorption to a great extent and osmotic absorption to a lesser extent. This would also pave the way to measure the growth of nations across the world and hence resulting in healthy, mutual talks amongst the leaders there by reducing conflicts overall.

EN 'd

www.ingramcontent.com/pod-product-compliance
Lightning Source LLC
Chambersburg PA
CBHW051019180526
45172CB00002B/406